EARLY AMERICAN SCIENTIFIC INSTRUMENTS and Their Makers

Frontispiece.—"Washington as a Surveyor." Engraving reproduced from Washington Irving's *Life of George Washington* (New York: 1857, vol. 1).

PLAN of the CITY of Washington in the Territory of Columbia,

ceded by the States of

VIRGINIA and MARYLAND

to the

United States of America,

and by them established as the

SEAT of their GOVERNMENT,

after the Year

MDCCC.

Engrav'd by Thackara & Vallance Philada. 1792.

Breadth of the Streets.

THE grand Avenues, and such Streets as lead immediately to public places, are from 130 to 160 feet wide, and may be conveniently divided into foot ways, walks of trees, and a carriage way. The other Streets are from 90 to 110 feet wide.

IN order to execute this plan, Mr. ELLICOTT drew a true Meridional line by celestial observation, which passes through the Area intended for the Capitol; this line he crossed by another due East and West, which passes through the same Area. These lines were accurately measured, and made the basis on which the whole plan was executed. He ran all the lines by a Transit Instrument, and determined the Acute Angles by actual measurement, and left nothing to the uncertainty of the Compass.

EARLY AMERICAN SCIENTIFIC INSTRUMENTS
and Their Makers

by

Silvio A. Bedini

THIS PUBLICATION ORIGINALLY APPEARED AS UNITED STATES NATIONAL MUSEUM BULLETIN 231, AND IS REPRINTED BY PERMISSION OF SMITHSONIAN INSTITUTION, WASHINGTON, D.C.

LANDMARK ENTERPRISES
10324 Newton Way
Rancho Cordova, CA 95670

ISBN 0-910845-30-1

Contents

Contents – Continued

Acknowledgments

The writer wishes to acknowledge his great indebtedness to the various compilations relating to clockmakers and instruments which have been consulted in the preparation of this work, and which have provided an invaluable basis for it.

He is especially grateful for the generous and interested assistance of the many who have cooperated in making this work possible. Particular credit must be given to Mrs. H. Ropes Cabot of the Bostonian Society; Mrs. Mary W. Phillips of the Department of Science and Technology of the U.S. National Museum; Prof. Derek J. de Solla Price, Avalon Professor of the History of Science at Yale University; Mr. Stephen T. Riley, Director of the Massachusetts Historical Society; and Mr. Charles E. Smart of Troy, New York.

'POINT OF BEGINNING'

When this work was first published more than twenty years ago, American scientific nstruments were not yet among the desiderata of collectors of Americana. There was, however, ome market in the United States for fine instruments by English and European makers. These vere generally sought by museum curators for displays to illustrate the history of science, and by a imited number of discerning scholarly collectors to implement their libraries. Regrettably, they lso became prizes for fashionable decorators in large cities seeking objects of erudition to serve s conversation pieces in the homes and offices of titans of commerce and industry.

Generally American instruments were of little interest, for they were not elaborately ecorated and engraved as were the foreign counterparts, having originally been intended rimarily for function and not to grace the cabinets of curiosities of princes, prelates and the vealthy. Generally they could not compete in grace and beauty with foreign instruments for everal reasons. Basically the American craftsmen were inclined to simple designs featuring the unctional over the decorative with cost to the purchaser as a major consideration. Until after the rst quarter of the nineteenth century all brass had to be imported, and was extremely expensive, o that brass instruments, whether imported from England or France, were often beyond the neans of the local surveyor, navigator or science teacher. Instruments produced in brass in the merican colonies and the new republic, chiefly by silversmiths and clockmakers, were also ecessarily substantially priced. These artisans advertised widely for used brass and copper which ould be reused by the process of cementation but still the price remained a factor. Finally the ntrepid Yankee whittlers of New England found a solution by producing surveying instruments arved or turned in wood utilizing the engraved compass cards used for marine compasses. These ould be made and sold for a fraction of the price of brass instruments, and proved to be quite as seful in the field. For the most part they were neither attractive nor decorative, and after the nid-nineteenth century when more sophisticated instruments became widely available, they were ast aside and forgotten. That they survived is due largely to their lack of attraction for interior ecorators for they were unsuitable for making into lamps or other "decorators." It was not until ne publication of the present work in 1964 that collectors and dealers began to recognize the ignificance of wooden surveying instruments, and they promptly jumped in price from several ollars to three figures.

Not all American instruments were simple in design, however. Surveying compasses made y Benjamin Rittenhouse, Goldsmith Chandlee, and Peregrine White among others, compare avorably in decorative elements with fine English instruments, undoubtedly because they relied n the same design books commonly in use among English makers. Today they are among the nost highly prized by modern collectors.

Inevitably the past two decades have seen the escalation of American instruments not only in nterest but also in price. An interesting observation is that among the most avid collectors of urveying instruments are practising civil engineers and land surveyors, who collect these relics as part of the history of their profession. A profession which although not the oldest in the world, is nonetheless one of the most respectable and as old as the nation itself, for it was the surveyors who cleared the wilderness for the first settlements and established communities and territorial boundaries. Fine examples of American-made instruments have become scarcer and scarcer as many have found their way into museums and the number of collectors has increased.

Within the past two decades, therefore, American instruments have come into their own. The fine antique English and European instruments have become infinitely scarce and rarely available. As a consequence American collectors have come to the realization that in the tools of the early American mathematical practitioners resided a part of their national history, that they were in fact the hardware of history.

At about the same time that *EARLY AMERICAN SCIENTIFIC INSTRUMENTS* appeared, several other works were published which proved useful for identifying surveying and navigational instruments, such as the late Charles E. Smart's *THE MAKERS OF SURVEYING INSTRUMENTS IN AMERICA SINCE 1700*, and the late Marion V. Brewington's catalogue of *THE PEABODY MUSEUM'S COLLECTION OF NAVIGATING INSTRUMENTS*. The catalogues of a number of dealers specializing in scientific instruments in modern times also have served as supplements to the existing literature.

Research undertaken and published subsequent to the appearance of the present volume has brought to light additional information about many of the individuals mentioned. Separate works have been published about Benjamin Banneker and Anthony Lamb, for example. These and others are considered as well in *Thinkers and Tinkers* (See *Addenda to Bibliography*).

The need to produce a comprehensive work on early American instrument makers and their surviving instruments has increased with the passing years and can now be fulfilled much more easily than would have been possible when this book first appeared. It remains a priority for the future.

Silvio A. Bedini
May, 1986

Preface

Within recent years fairly exhaustive studies have been made on many aspects of American science and technology. For example, there have been numerous works relating to clocks and clockmakers, so that the collector and horological student have a number of useful sources on which to rely. More recently there has been a series of publications on the development of American tools and their makers. Until now, however, no systematic study has been attempted of the scientific instruments used in the United States from its colonial beginnings. While several useful regional lists of instrument makers in early America have been compiled from advertisements in contemporary newspapers and published as short articles, these, however, are fragmentary, and are inadequate to the need for documentation in this field.

With the rapidly growing interest in the history of science, it becomes necessary to have a more complete background for the student and the historian alike. It is desirable to have a more comprehensive picture of the work of the scientific practitioners of the earlier periods of American scientific development, and of their tools. At the same time it is essential to have a history of the development and distribution and use of scientific instruments by others than the practitioners and teachers. The role of the instrument maker in the American Colonies was an important one—as it was in each epoch of the history of science in Europe—and it deserves to be reported.

To make a comprehensive study of American scientific instruments and instrument makers in the American Colonies is no simple matter, partly because of an indifference to the subject in the past, and partly because of the great volume of sources that must be sifted to accomplish it. Such a project would require an organized search of all published reference works relating to the field and associated topics, of all contemporary newspapers for advertisements and notices, of civil records filed in state and community archives, of business account-books and records that have been preserved, and of business directories of the period under consideration. In addition, such a study would require the compilation of an inventory of all surviving instruments in private and public collections, and a correlation of all the data that could be assembled from these sources.

The present study attempts only in part to accomplish this aim, being no more than a preliminary compilation of the scientific instruments known to have been used during the first two centuries of American colonial existence. It merely attempts to assemble all the data that is presently available in scattered sources, and to organize it in a usable form for the student and historian of American science. A supplement relating to 19th-century instruments and instrument makers is in progress.

The most that is hoped for the present work is that it will be of temporary assistance, serving to bring forth additional information on the subject from sources not previously available or known.

February 1, 1964 S.A.B.

EARLY AMERICAN SCIENTIFIC INSTRUMENTS
and Their Makers

by

Silvio A. Bedini

The Tools of Science

Philosophical and Practical Instruments

DEVELOPMENT OF THE SCIENCES in the American Colonies was critically dependent upon the available tools—scientific instruments—and the men who made and used them. These tools may be separated into two groups. The first group consists of philosophical instruments and scientific teaching apparatus produced and employed for experimentation and teaching in educational institutions. The second includes the so-called "mathematical instruments" of practical use, which were employed by mathematical practitioners and laymen alike for the mensural and nautical needs of the Colonies. It is particularly with this second group that the present study is concerned.

It has been generally assumed that scientific instruments, as well as the instrument makers, of the first two centuries of American colonization were imported from England, and that the movement declined by the beginning of the 19th century with the development of skilled native craftsmen.[1] This assumption is basically true for those instruments grouped under philosophical and scientific apparatus for experimentation and teaching. Almost all of these items were in fact imported from England and France until well into the 19th century.

Likewise, the very earliest examples of mathematical instruments for surveying and navigation in the Colonies were imported with the settlers from England. It was not long after the establishment of the first settlements, however, that the settlers, and later the first generation of native Americans, began to produce their own instruments. Records derived from historical archives and from the instruments themselves reveal that a considerable number of the instruments available and used in the Colonies before 1800 were of native production. Apparently, relatively few instrument makers immigrated to the American continent before the end of the

[1] DEREK J. DE SOLLA PRICE, *Science Since Babylon* (New Haven: Yale University Press, 1961), pp. 62–64.

Revolutionary War. Later, with the beginning of the 19th century, makers of and dealers in instruments in England and France became aware of the growing new market, and emigrated in numbers to establish shops in the major cities of commerce in the United States.

Quite possibly the few instrument makers trained in England who immigrated to the Colonies in the early epoch of Colonial development may have in turn trained others in their communities, although no evidence has yet been found. Perhaps more data on this aspect of the subject will eventually come to light.

There is reason to believe that a few mathematical practitioners and instrument makers lived and worked in the New England colonies as early as the first century of colonization.

The evidence, frankly meager, consists of two items. The first is a reference relating to James Halsie of Boston. In a land deed made out to him in 1674 he was referred to as a "Mathematician."[2] Halsie was listed as a freeman of the Massachusetts Bay Colony in 1690. He apparently was the forbear of the several members of the Halsy family of instrument makers of Boston of the 18th century, mentioned later in this study. It is uncertain whether the use of the term "mathematician" in this connection meant an artisan, but if not it may be inferred that Halsie was a practitioner.

The second piece of evidence is even more slender; it consists of an inscription upon a dialing rule (fig. 1) for making sundials and charts. The instrument is of cast brass, 20⁷⁄₁₆ inches long and 1¹¹⁄₁₆ inches wide. The date "1674" is inscribed on the rule together with the name of its original owner, "Arthur Willis." The instrument almost certainly was produced by the school of Henry Sutton, the notable English instrument maker who worked in Threadneedle Street in London from about 1637 through 1665. The name and date inscriptions are consistent and contemporary with the workmanship of the rule, and were probably inscribed by the maker for the original owner. It is conceivable that Arthur Willis was an Englishman and that the rule was brought into this country even in relatively recent times. However, it is claimed that the rule was owned and used by Nathaniel Footes, surveyor of Springfield, Massachusetts. Nathaniel Footes, believed to have been originally from Salem, subsequently moved from Springfield to Wethersfield, Conn. The instrument was later owned and used

---

[2] James Savage, *A Genealogical Dictionary of the First Settlers of New England* (Boston, 1860), vol. 2, p. 341.

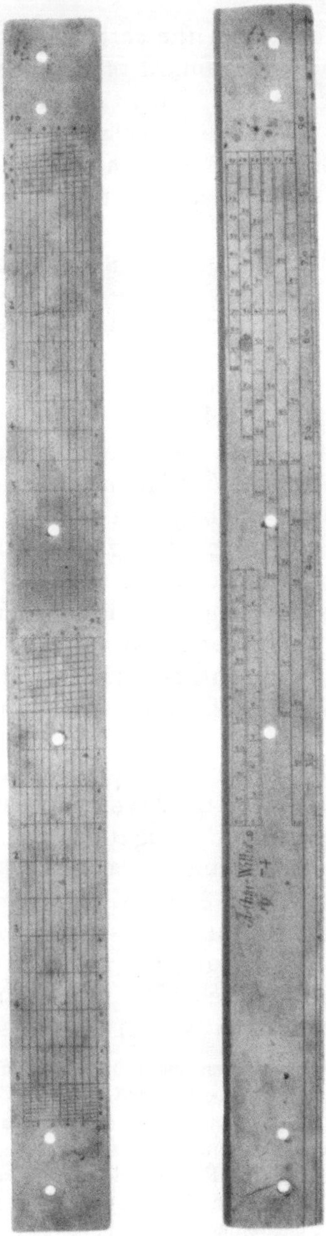

Figure 1.—Dialing rule made of brass and inscribed with the name "Arthur Willis" and the date "1674." Allegedly used by Nathaniel Footes, surveyor of Springfield, Massachusetts. Photo courtesy Newton C. Brainard, Hartford, Connecticut, and the Connecticut Historical Society.

in Connecticut not later than the early 19th century [3] by the forbears of Mr. Newton C. Brainard of Hartford, Connecticut. If records relating to Willis as a resident of the New England colonies can be recovered, it may then be possible to establish whether he worked in the Colonies as a mathematical practitioner in the 17th century. His name is included on a tentative basis.

The Need for Instruments

The production and use of scientific instruments in the American Colonies reflected colonial development in education and in territorial and economic expansion, and closely paralleled the same development in England, where the first mathematical practitioners were the teachers of navigational and commercial arithmetic and the surveyors employed in the redistribution of land following the dissolution of the monasteries. As the communities became established and the settlers gained a foothold on the soil, their attention naturally turned to improving their lot by expanding the land under cultivation and by trading their products for other needs. The growth of the communities became increasingly rapid from the end of the 17th century, and the land expansion closely paralleled the development of trade. The educational institutions placed greater emphasis on the sciences as their curriculums developed. Particularly there was a greater preoccupation with the sciences on the part of the layman because of the need for knowledge of surveying and navigation.

The colonial school curriculum was accordingly designed from the practical point of view to emphasize practical mathematics, and there was an increasing demand for instruction in all aspects of the subject. One of the earliest advertisements of this nature appeared in *The Boston Gazette* in March 1719. In the issue of February 19 to March 7 the advertisement stated that:

> This day Mr. Samuel Grainger opens his school at the House formerly Sir Charles Hobby's, where will be taught Grammar Writing after a free and easy manner in all the usual Hands, Arithmetick in a concise and Practical Method, Merchants Accompts, and the Mathematicks.
>
> He hopes that more thinking People will in no wise be discouraged from sending their children thither, on the account of the reports newly reviv'd, because these dancing Phaenomena's were never seen nor heard of in School Hours.

---

[3] *The Chronicle* (Early American Industries Association), March 1936, vol. 1, no. 16, p. 8; and personal correspondence with Mr. William L. Warren, Connecticut Historical Society.

The advertisement was further amplified in its second appearance, in the issue of March 21–22, 1719:

> At the house formerly Sir Charles Hobby's are taught grammar, writing, after a free & easy manner in all hands usually practiced, Arithmetick Vulgar and Decimal in a concise and Practical Method, Merchants Accompts, Geometry, Algebra, Mensuration, Geography, Trigonometry, Astronomy, Navigation and other parts of the Mathematicks, with the use of the Globes and other Mathematical Instruments, by Samuel Grainger.
>
> They whose business won't permit 'em to attend the usual School Hours, shall be carefully attended and Instructed in the Evenings.

R. F. Seybold [4] has noted that: "In advertisements of 1753 and 1754, John Lewis, of New York City, announced 'What is called a New Method of Navigation, is an excellent Method of Trigonometry here particularly applied to Navigation; But it is of great use in all kinds of measuring and in solving many Arithmetical Questions.' James Cosgrove, of Philadelphia, in 1755, taught 'geometry, trigonometry, and their application in surveying, navigation, etc.,' and Alexander Power, in 1766, 'With their Application to Surveying, Navigation, Geography, and Astronomy'." These subjects were featured also in the evening schools of the colonial period, maintained by private schoolmasters in some of the larger communities for the education of those who could not attend school in the daytime.

According to Seybold, surveying and navigation were the most popular mathematical subjects taught. Some explanation is to be derived from the statement by Schoen [5] that: "In the days when the 'bounds' of great wilderness tracts were being marked off by deep-cut blazes in the trees along a line, a knowledge of land surveying was a useful skill, and many a boy learned its elements by following the 'boundsgoer' in his work of 'running the line.' And those who did not actually take part in running the line must have attended many a gay springtime 'processioning' when neighbors made a festive occasion out of 'perambulating the bounds'. "Vague land grants and inaccurate surveys," he adds, "made the subject of boundary lines a prime issue in the everyday life of colonial homes."

---

[4] R. F. Seybold, "The Evening School in Colonial America," *Bureau of Educational Research, Bulletin 31* (University of Illinois, 1925), p. 28.

[5] H. H. Schoen, "The Making of Maps and Charts," *Ninth Yearbook of the Council for the Social Studies* (Cambridge, 1938), p. 83; also Edmond R. Kiely, *Surveying Instruments: Their History and Classroom Use* (New York: Teachers College, Columbia University, 1947), pp. 239–250.

At the same time there was interest in the other aspects of the mathematical sciences. As early as 1743, for instance, a Harvard mathematician named Nathan Prince advertised in Boston that if he were given "suitable Encouragement" he would establish a school to teach "Geography and Astronomy, With the Use of the Globes, and the several kinds of Projecting the Sphere" among other things.[6] A decade later, Theophilus Grew, professor in the academy at Philadelphia which has become the University of Pennsylvania, published a treatise on globes, with the title:

> *The Description* and *Use* of the *Globes*, Celestial and Terrestrial; With Variety for *Examples* for the Learner's *Exercises*: Intended for the Use of Such Persons as would attain to the Knowledge of those *Instruments*; But Chiefly designed for the *Instruction* of the young *Gentlemen* at the *Academy* in Philadelphia. To which is added Rules for working all the Cases in Plain and Spherical Triangles without a Scheme. By *Theophilus Grew*, Mathematical Professor. Germantown, Printed by Christopher Sower, 1753.[7]

Thus, the need for practical mathematical instruments for the surveyor and navigator became critical in proportion to the need for men to make and use them, and it is not surprising to discover that the majority of the instruments produced and advertised by early American makers were for surveying, with nautical instruments in second place. Generally, the surveyors were not professionals; they were farmers, tradesmen, or craftsmen with a sound knowledge of basic arithmetic and occasionally with some advanced study of the subject as taught in the evening schools. The surveying of provincial and intercolonial boundaries required greater skill, however, as well as a knowledge of astronomy, and this work was relegated to the scientific men of the period.

As the increasing preoccupation with subdivision of land and with surveying led to a greater demand for suitable instruments, it was the skilled craftsmen of the community, such as the clockmaker and the silversmith, that were called upon to produce them. Superb examples also were produced by the advanced scientific men, or "mathematical practitioners," of the period.

Colonial Training in Instrument Making

One may well ask, where did these native craftsmen acquire the knowledge that enabled them to produce so skillfully the accurate

[6] Brooke Hindle, *The Pursuit of Science in Revolutionary America 1735–1789* (Chapel Hill, N.C.: University of North Carolina Press, 1956), pp. 337–338.

[7] LeRoy E. Kimball, "James Wilson of Vermont, America's First Globe Maker," *Proceedings of the American Antiquarian Society* (April 1938), p. 31.

Figure 2.—Title page of *The Surveyor* by Aaron Rathborne, published in London in 1616. The book was one of the sources of information for American makers of mathematical instruments.

and often delicate mathematical instruments? There were a number of possible sources for this knowledge. The first source lies in England, where some of these craftsmen could have studied or served apprenticeships. After completing their apprenticeship with English mathematical practitioners, they may have immigrated to the Colonies and taught the craft to others. This seems to be entirely plausible, and was probably true, for example, of Thomas Harland the clockmaker, Anthony Lamb, and perhaps several others. However, these were the exceptions instead of the rule, since a biographical study of the instrument makers in general reveals that they were for the most part native to America. It is not likely that the one or two isolated practitioners that had been trained in England could have taught so many others who worked in the same epoch.

Another source for this knowledge of instrument making was probably the reference works on the subject that had been published in England and in France. As an example, Nicolas Bion's *Traite de la Construction et des Principaux Usages des Instruments de Mathematique*, which had been first published in 1686, was translated into English by Edmund Stone in 1723, and went into several English editions. Copies of this work in English undoubtedly found their way to America soon after publication. Other popular works were Aaron Rathbone's *The Surveyor*, which appeared in London in 1616 (see fig. 2); William Leybourn's *The Compleat Surveyor*, in 1653; and George Atwell's *Faithfull Surveyour*, in 1662. Other works popular in the Colonies were R. Norwood's *Epitome, or The Doctrine of Triangles* (London, 1659) and J. Love's *Geodasia, or the Art of Surveying* (London, 1688).

These works undoubtedly inspired similar publications in America, for many books on surveying and navigation appeared there before the beginning of the 19th century. Chief among them were S. Moore's *An Accurate System of Surveying* (Litchfield, Conn., 1796), Z. Jess's *A Compendious System of Practical Surveying* (Wilmington, 1799), Abel Flint's *Surveying* (Hartford, 1804), and J. Day's *Principles of Navigation and Surveying* (New Haven, 1817).

The published works were unquestionably responsible for much of the training in the making of mathematical instruments in America, although no documentary evidence has yet been recovered to prove it.

Another important influence on early American instrument-making which must be noted was that of the clockmaker as an artisan. A comprehensive study of surviving instruments and

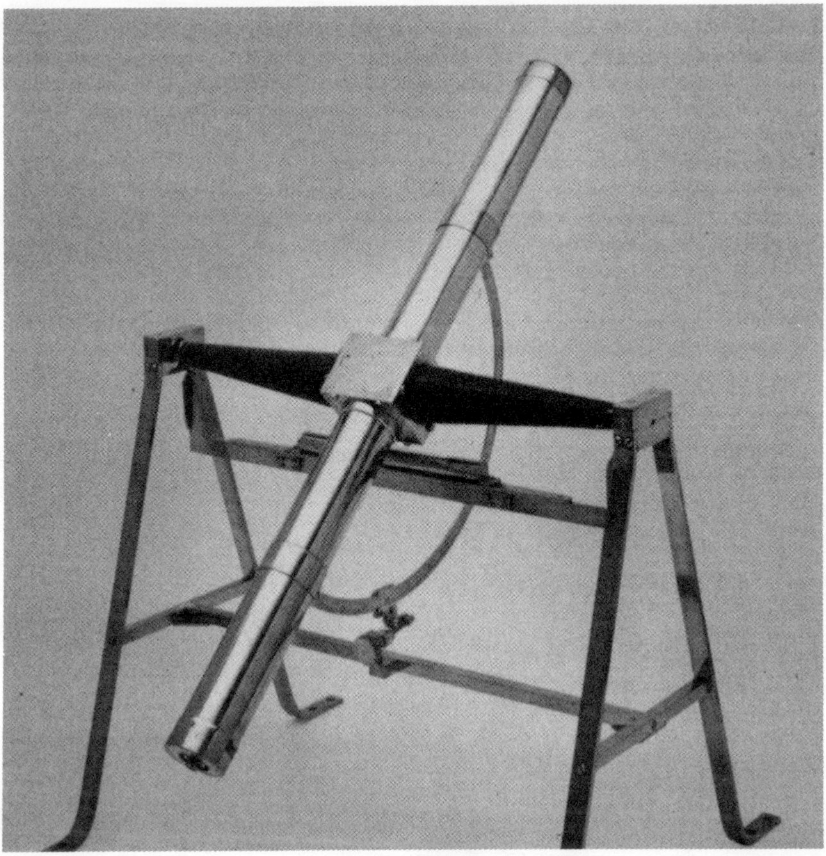

Figure 3.—Transit telescope made by David Rittenhouse and used by him for the observation of the transit of Venus in 1769. Brass, 33½-in. tube on a 25-in. axis, with an aperture of 1¾ in. and a focal length of 32 in. Photo courtesy the American Philosophical Society.

related records has revealed that only a few of the many clockmakers working in the American Colonies in the 18th century made mathematical instruments. Yet, a large proportion of the surviving surveying and nautical instruments produced before 1800 were the work of clockmakers. Classic among these must be noted the instruments produced by the brothers David and Benjamin Rittenhouse (see p. 15 and figs. 3 and 4), as well as the fine surveying instruments made by four separate members of the Chandlee family, whose clockmaking traditions began early in the 17th century (see p. 54).

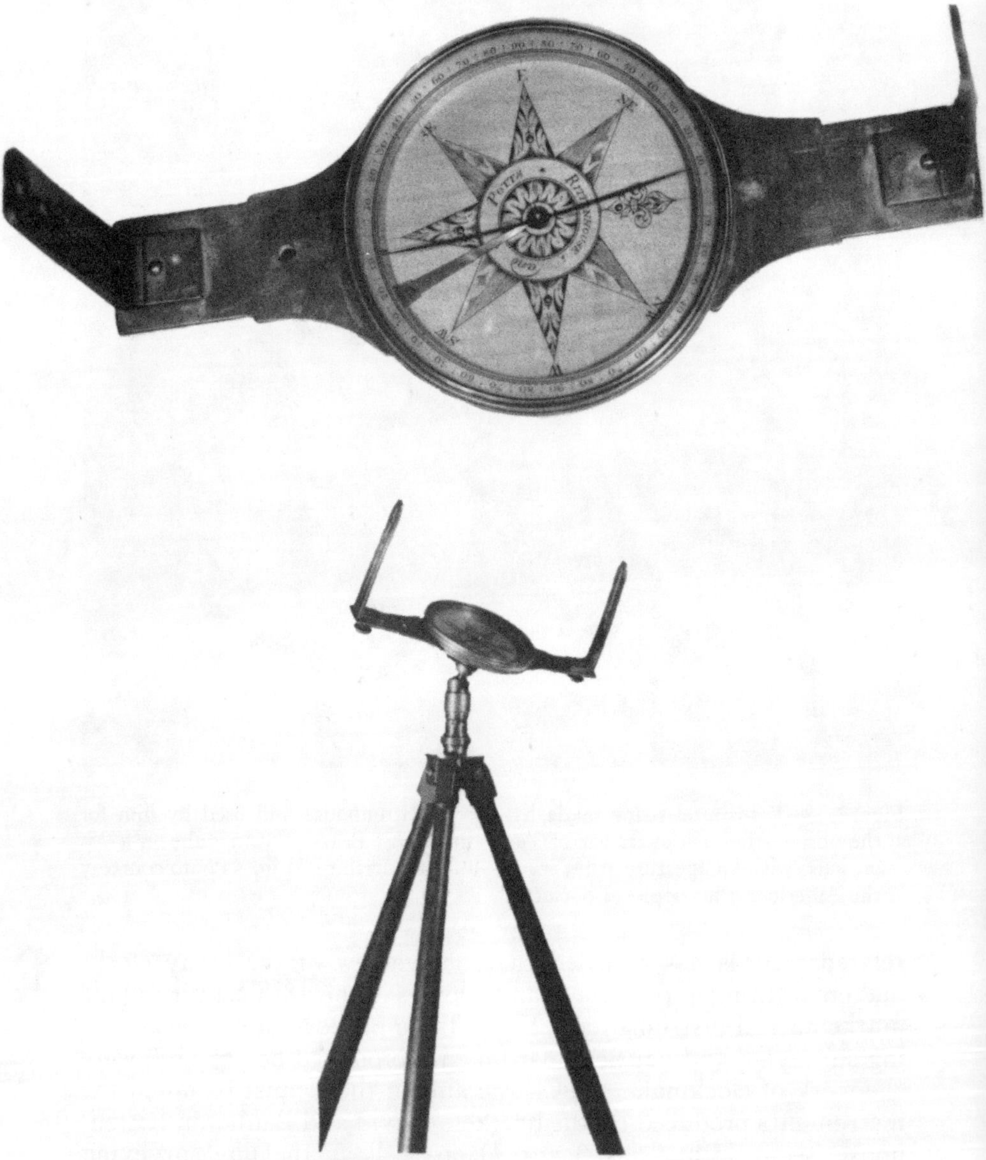

Figure 4.—Surveying compass marked "Potts and Rittenhouse." Believed to be the work of David Rittenhouse in partnership with Thomas Potts. Photo courtesy the American Philosophical Society.

Finally, one must not overlook the fact that examples of English and other European instruments were available in the Colonies, and that at least some of the early colonial makers undoubtedly copied them. It is apparent from some surviving early American instruments that the materials, designs, dimensions, and details of European prototypes had been deliberately copied. It is possible to see in public collections, for instance, a Davis quadrant of English manufacture exhibited beside a later example, signed by a New England maker, which comes extraordinarily close to duplicating it in every feature.

As with the presumed influence of published works, the practice of copying imported instruments cannot be documented, but it must have been engaged in by many of the unschooled New England instrument makers. By this means some may even have profited to the degree that they became professional craftsmen without benefit of formal apprenticeship.

Yet it is remarkable that although numerous instruments were produced by native artisans, in addition to the substantial number which were imported before the end of the 18th century, relatively few specimens have survived in public collections as well as in private hands. Despite the exhaustive combing of attics and barns throughout the country by dealers in antiques and by avid collectors during the past several decades, the number of surviving instruments now known is incredibly small in comparison with the numbers known to have been made locally or imported before the beginning of the 19th century. Since instruments are not items which would ordinarily be deliberately discarded or destroyed, or melted down for the recovery of the metal, this small percentage of survival presents a puzzle which has not been resolved.

Figure 5.—David Rittenhouse. Engraving from portrait by Charles Wilson Peale.

The Mathematical Practitioners

The Rittenhouse Brothers

NOTABLE AMONG THE American practitioners was David Rittenhouse (1732–1796) of Norristown and Philadelphia, Pennsylvania, who was established as a clockmaker and surveyor in Philadelphia by 1749. He surveyed the boundary between Pennsylvania and Delaware in 1763 with instruments of his own design and construction. Six years later, in 1769, he successfully calculated the transit of Venus and later observed that planet with astronomical instruments he had constructed himself. In the following year, 1770, he built the first American astronomical observatory, in Philadelphia. Two orreries that he designed and built—at the University of Pennsylvania and at Princeton University—survive as outstanding examples of American craftsmanship.[8] Several of his surveying and astronomical instruments are exhibited in the collections of the U.S. National Museum. David Rittenhouse is credited with being the originator of a declination arc on the surveying compass, a feature to be copied by a number of later instrument makers.

David's brother, Benjamin Rittenhouse (1740–c.1820), served in the Revolution and was wounded at Brandywine. He superintended the Government's gunlock factory at Philadelphia in 1778 and achieved recognition as a maker of clocks and surveying instruments (see fig. 8).[9] During one period of his career he worked in partnership with his brother David. An interesting advertisement appeared in the May 14, 1785, issue of *The Pennsylvania Packet:*

> WANTED, An ingenious Lad not exceeding 14 years of age, of a reputable family, as an Apprentice to learn the Art and Mistery of making Clocks and Surveying Instruments. Any lad inclining to go an apprentice to the above Trade, the terms on which he will be taken may [be] known by enquiring of Mr. David Rittenhouse, in Philadelphia, or at the subscriber's house in Worcester township, Montgomery county. Benjamin Rittenhouse.

---

[8] HINDLE, op. cit. (footnote 6).

[9] GEORGE H. ECKHARDT, *Pennsylvania Clocks and Clockmakers* (New York: Devin-Adair Co., 1955), p. 190.

Figure 6.—Astronomical clock made by David Rittenhouse for his observatory at Norristown, Pa., and used by him for the observation of the transit of Venus in 1769. Unembellished pine case 83½ in. high, 13¼ in. wide at the waist with a silvered brass dial 10⅝ in. diameter. Photo courtesy the American Philosophical Society.

Figure 7.—Orrery built by David Rittenhouse for the University of Pennsylvania. The center section shows the motions of the planets and their satellites and the right-hand section the eclipses of the Sun and Moon. The case, considered to be an outstanding example of colonial cabinetwork, was make by John Folwell.

Figure 8.—Brass surveying compass inscribed "Made by Benjamin Rittenhouse, 1787." Photo courtesy Ohio State Museum, Columbus, Ohio.

FIGURE 7

FIGURE 8

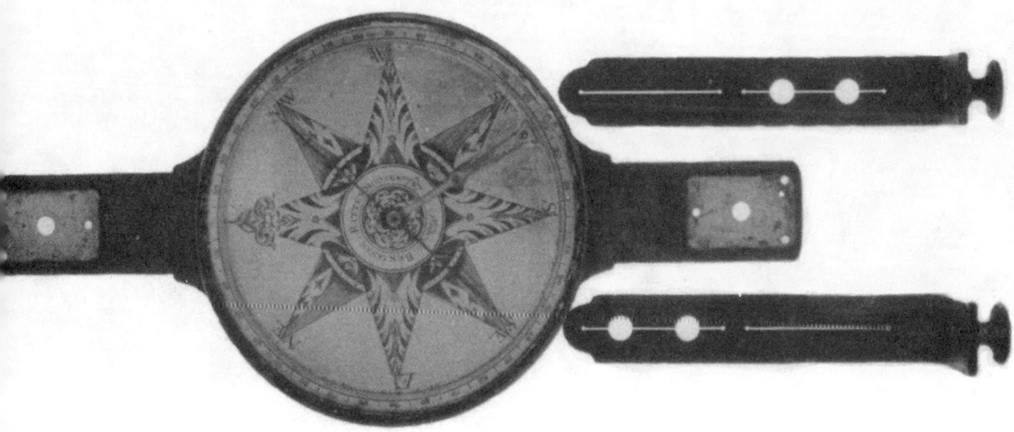

Figure 9.—Portrait of Andrew Ellicott (1754–1820) by unknown artist.

Andrew Ellicott

A name closely associated with that of the Rittenhouse brothers was that of Andrew Ellicott (1754–1820) of Solebury, Pennsylvania, and Ellicotts Mills, Maryland. Andrew was the son of Joseph Ellicott, the clockmaker and pioneer industrialist who founded Ellicotts Mills. Although a Quaker, Andrew (fig. 9) served in the Revolution, and he became one of the most distinguished engineers of the new republic. He worked as a clockmaker and instrument maker from 1774 to 1780. In 1784 he ran the boundary between Virginia and Pennsylvania and in the following year he was a member of the survey that continued Mason and Dixon's line. In 1785 and 1786 he served on the Pennsylvania commissions that surveyed the western and northern boundaries of the state, and in 1789 he served on the commission that fixed the boundary between New York and Pennsylvania. Between 1791 and 1793 he surveyed the site of the city of Washington, D.C., and redrew L'Enfant's plan for the city.

In early 1793 Ellicott was appointed commissioner by the Commonwealth of Pennsylvania for the project of viewing and locating a road from Reading to Presque Isle, now Erie. It was an extremely difficult undertaking, but Ellicott completed the work by the autumn of 1796, including laying out the towns of Erie, Warren, and Franklin.

In May 1796 Ellicott was commissioned by President Washington to survey and mark the boundary line between the United States and the Spanish Province of Florida in accordance with the provisions of the Pinkney-Godoy Treaty of October 27, 1795. This line was to begin at the point where the 31st parallel of north latitude intersected the Mississippi River, and to proceed thence along that parallel eastward to the Appalachicola River for about 400 miles.

In 1801 Ellicott was offered the position of surveyor general of the United States by President Jefferson. Ellicott declined, but subsequently accepted the secretaryship of the land office of Pennsylvania, a post he held until 1808.

In 1811 Ellicott became commissioner to represent Georgia in locating the Georgia-North Carolina boundary, a project on which he was engaged for the major part of the following year.

In 1815 President Madison appointed Ellicott professor of mathematics at West Point, with the rank of major. This is an appointment he kept until his death in 1820. It was interrupted

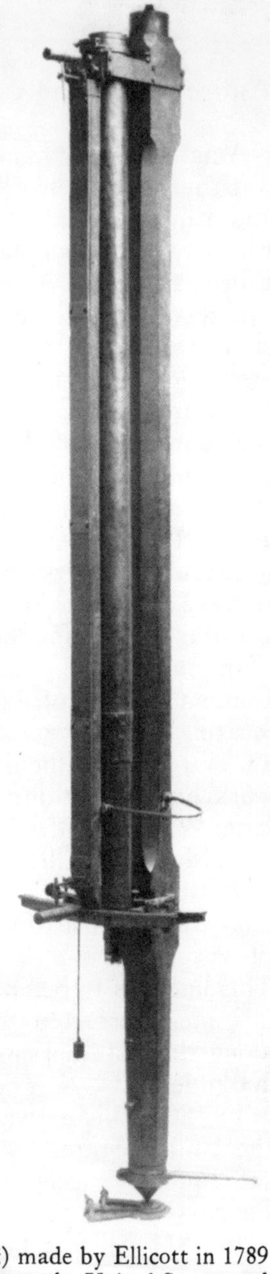

Figure 10.—Transit and equal-altitude instrument (left) made by Ellicott in 1789 and used by him in the survey of the boundary between the United States and Florida and in other surveys. USNM 152080.

Figure 11.—Zenith sector with focal length of 6 ft., made by David Rittenhouse and revised by Andrew Ellicott. Described in *Journal of Andrew Ellicott* (Philadelphia, 1803). USNM 152078.

in 1817 when the Government required his services as astronomer to locate a portion of the United States-Canadian boundary in accordance with the fifth article of the Treaty of Ghent.

Ellicott was a member of a number of learned societies, including the American Philosophical Society, the Society for the Promotion of Useful Arts of Albany, and of the National Institute of France.

Ellicott constructed a number of instruments for surveying and astronomical observation, and he designed and used others that were produced by his friend David Rittenhouse [10] (see figs. 10, 11). Of particular interest in connection with Ellicott's career as a clockmaker and instrument maker are two advertisements that appeared in the Baltimore newspapers. The first one was in the *Maryland Journal and Baltimore Daily Advertiser on April 7, 1778:*

> Ellicott's Upper Mills, April 4, 1778. Wanted, a person acquainted with the Clock-Making business, and able to work by directions. Such a person will meet with good encouragement by applying to Andrew Ellicott, sen.

The second advertisement, in the same vein, appeared in the May 16, 1780, issue of the *Maryland Journal:*

> Good Encouragement will be given to either Clock or Mathematical instrument makers, by the subscriber, living in Baltimore-Town. Andrew Ellicott.

Owen Biddle

Another mathematical practitioner associated with David Rittenhouse in his observations of the transit of Venus was Owen Biddle (1737–1799) of the North Ward, Philadelphia.

In early life Biddle was an apothecary and a clock- and watchmaker. In his shop "next door to Roberts warehouse" he sold clock and watch parts and tools. From 1764 to 1770 he advertised himself as "Clockmaker, and scientist, statesman and patriot." As a Quaker, he participated actively in civic and patriotic affairs of Philadelphia. During the American Revolution, in spite of his religious affiliation, he fought for the defense of the Colonies and was appointed Deputy Commissioner of Forage. Evidencing sincere repentance, he was permitted to rejoin the Society of Friends.

In 1769 Biddle took an active part in the preparations made by the American Philosophical Society for the observation of the transit of Venus. With Joel Baily he was sent to Cape Henlopen, Delaware, with a large reflecting telescope borrowed from the Li-

[10] Catherine Van C. Mathews, *Andrew Ellicott, His Life and Letters* (New York, 1908).

brary Company. The expedition was described in the *Transactions of the American Philosophical Society* in 1771 in an article entitled "An Account of the Transit of Venus, over the Sun's Disk, as observed near Cape Henlopen, on Delaware Bay, June 3rd, 1769 by Owen Biddle, Joel Baily and (Richard Thomas) Drawn by Owen Biddle." In addition to his trade in clocks and watches, Biddle also made mathematical instruments and was well known in his native city as a merchant, inventor, and ironmaster.

Benjamin Banneker

A name that is too often ignored in the history of science in colonial America is that of a free Negro, Benjamin Banneker (c. 1734–1806) of Baltimore. A farmer by occupation, Banneker was the son of a native African slave and a free mulatto woman. In his spare time he attended the school of a Quaker farmer; the only book he owned was the Bible. When he was a young man he acquired a watch from a trader, and from it he developed his love of science and instruments. Although he had never seen a clock, he constructed one based on drawings he made from the watch. Banneker was called upon to assist in the construction of the mills for the Ellicotts, and it was natural that his clock, which was the marvel of the Negro settlement, should come to the attention of Joseph Ellicott. Ellicott became interested in Banneker's thirst for knowledge and allowed him the use of his tools, scientific instruments, and technical books. Among the books were Mayer's *Tables*, James Ferguson's *Astronomy*, and Leadbeater's *Lunar Tables*. Banneker absorbed these and other works that he borrowed and went on to explore the wonderful new world they opened up for him. He pursued astronomical studies, and within three years he began to make calculations (fig. 12) for an almanac. After completing the calculations for the year 1791, he went on to produce a set of calculations for 1792. During this period he mastered the use of surveying instruments and made a firsthand study of tides in the region. His great opportunity came when Andrew Ellicott was chosen to make a survey for the city of Washington and hired Banneker as an assistant. While thus employed, Banneker completed his almanac and gave it to George Ellicott, Andrew's cousin, as a subject of possible interest. Apparently George Ellicott turned it over to the Honorable James McHenry of Baltimore, who in turned submitted it to the Philadelphia firm of Goddard & Angell, who published it (fig. 13). Banneker mailed

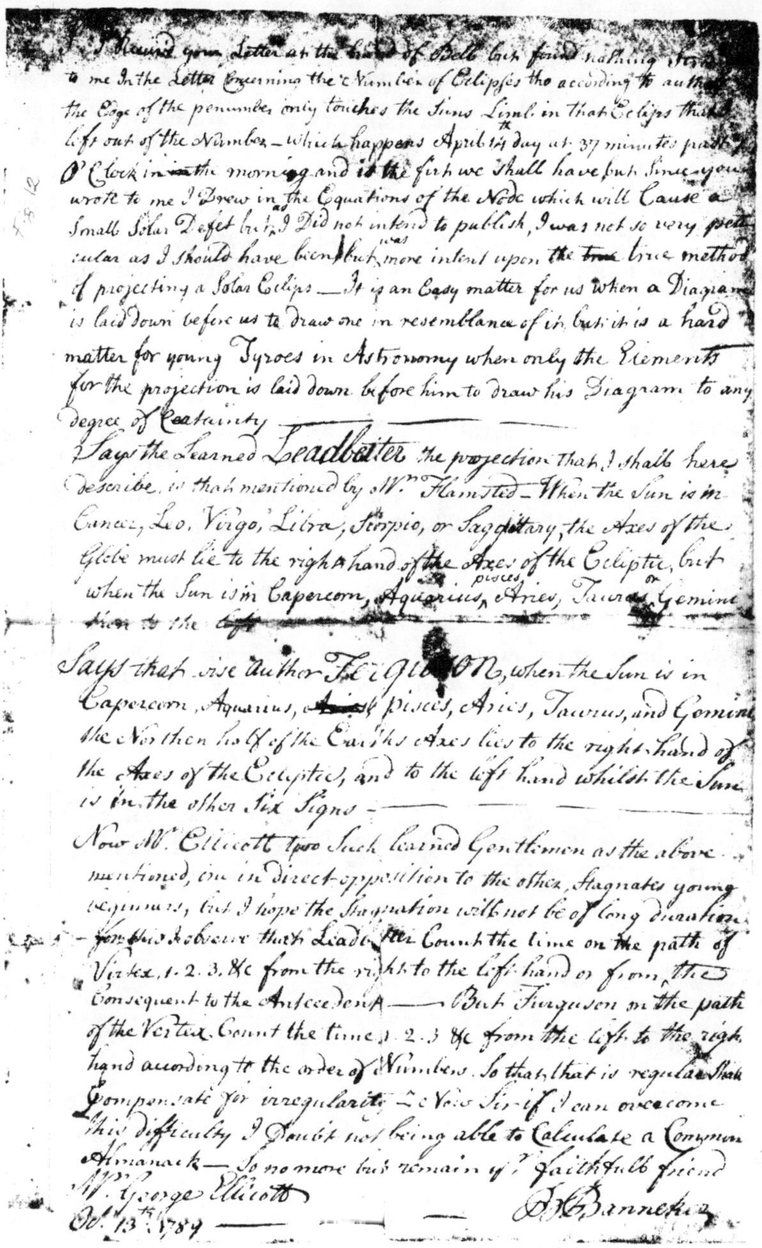

I Received your Letter at the hand of Bell but found nothing strange to me In the Letter Concerning the Number of Eclipses tho according to authors the Edge of the penumber only touches the Suns Limb in that Eclips that I left out of the Number—which happens April 14th day at 37 minutes past O'Clock in the morning and is the first we shall have but Since you wrote to me I Drew in the Equations of the Node which will Cause a Small Solar Defect but as I Did not intend to publish, I was not so very perticular as I should have been but was more intent upon the true method of projecting a Solar Eclips— It is an Easy matter for us when a Diagram is laid down before us to draw one in resemblance of it, but it is a hard matter for young Tyroes in Astronomy when only the Elements for the projection is laid down before him to draw his Diagram to any degree of Certainty—

Says the Learned Leadbetter the projection that I shall here describe, is that mentioned by Mr. Flamsted—When the Sun is in Cancer, Leo, Virgo, Libra, Scorpio, or Saggitary, the Axes of the Globe must lie to the right hand of the Axes of the Ecliptic, but when the Sun is in Capercorn, Aquarius, pisces, Aries, Taurus, or Gemini ~~then to the left~~

Says that wise Author Ferguson, when the Sun is in Capercorn, Aquarius, pisces, Aries, Taurus, and Gemini the Northen half of the Earths Axes lies to the right hand of the Axes of the Ecliptic, and to the left hand whilst the Sun is in the other Six Signs—

Now Mr. Ellicott two Such learned Gentlemen as the above mentioned, one in direct opposition to the other, stagnates young beginners, but I hope the Stagnation will not be of long duration—for this I observe that Leadbeter Count the time on the path of Vertex, 1. 2. 3. &c from the right to the left hand or from the Consequent to the Antecedent— But Ferguson on the path of the Vertex Count the time, 1. 2. 3 &c from the left to the right hand according to the order of Numbers. So that, that is regular shall Compensate for irregularity— Now Sir if I can overcome this difficulty I Doubt not being able to Calculate a Common Almanack— So no more but remain yr. faithfull friend

Mr. George Ellicott

Octr. 13th 1789

B. Banneker

Figure 12.—Letter from Benjamin Banneker to George Ellicott dated October 13, 1789, regarding astronomical data for the compilation of Banneker's almanac. Photo courtesy Maryland Historical Society.

a copy of his *Benjamin Banneker's Pennsylvania, Delaware, Virginia And Maryland Almanac and Ephemeris For the Year of Our Lord, 1792* to Thomas Jefferson, who was so impressed with it that he forwarded it to the Marquis de Condorcet, secretary of the French Academy of Sciences. After his work with Ellicott had been completed, Banneker retired to his farm to produce almanacs annually until 1802. When he died in 1806 he was eulogized before the French Academy by the Marquis de Condorcet, and William Pitt placed his name in the records of the English Parliament.[11]

Joel Baily

Still another 18th-century practitioner was Joel Baily (1732–1797), a Quaker of West Bradford, Pennsylvania. In addition to his trade as a clockmaker and gunsmith, Baily achieved local eminence as an astronomer, mathematician, and surveyor.[12]

In 1764, at the time that Charles Mason and Jeremiah Dixon established their headquarters near his farm, Baily was the local surveyor. Obtaining employment with the expedition, he worked with Mason and Dixon until the completion of their survey in 1768. Baily was subsequently employed by Mason and Dixon to build pine frames for carrying the 20-foot rods to be used in the second measurement of courses from the Stargazers' Stone southward.

In 1769 Baily was appointed by the American Philosophical Society to work with Owen Biddle in setting up the station at Cape Henlopen for observation of the transit of Venus. In 1770 he again worked with Biddle in taking the courses and distances from the New Castle Court House to the State House Observatory in Philadelphia for determining the latitude and longitude of each. In the same year Baily was elected a member of the American Philosophical Society.

Reverend John Prince

Another noteworthy mathematical practitioner of the period was the Reverend John Prince (1751–1836) of Salem, Massachusetts.

---

[11] John H. B. Latrobe, "Memoir of Benjamin Banneker," *Maryland Colonization Journal* (Baltimore, May 1845); Philip LePhillips, "The Negro, Benjamin Benneker," *Records of the Columbia Historical Society* (1916), vol. 20.

[12] Arthur E. James, *Chester County Clocks and Their Makers* (West Chester, Pa.: Chester Historical Society, 1947), pp. 29–39; *Transactions of the American Philosophical Society*, ser. 1, vol. 1, pp. 85–97.

Benjamin Bannaker's
PENNSYLVANIA, DELAWARE, MARY-
LAND, AND VIRGINIA
ALMANAC,
FOR THE
YEAR of our LORD 1795;
Being the Third after Leap-Year.

BANNAKER.

—PRINTED FOR—
And Sold by JOHN FISHER, *Stationer.*
BALTIMORE.

Figure 13.—Title page of one of Banneker's almanacs. The portrait of Banneker was made by Timothy Woods in 1793 for the publisher and reproduced by woodcut. Banneker's first almanac was published in Philadelphia in 1792.

The son of a hatter and mechanic, Prince studied natural philosophy under John Winthrop at Harvard and received his B.A. degree in 1776. He was a student of divinity under Samuel Williams and was ordained in 1779 at the First Church in Salem. Although an amateur of the sciences, Prince became a skilled maker of scientific instruments. He made, sold, and repaired instruments for the use of numerous colleges, schools, and academies, including Brown, Dartmouth, Rutgers, Harvard, Union, Amherst, and Williams. Among other accomplishments, he effected "improvements" on the lucernal microscope and the air pump.[13]

Amasa Holcomb

Although he was born in the 18th century, Amasa Holcomb (1787–1875) properly belongs to a later period. An astronomer and telescope maker of Southwick, Massachusetts, Holcomb became a surveyor in 1808. An autobiographical sketch noted that "he manufactured about this time a good many sets of surveyors instruments—compasses, chains, scales, protractors and dividers, some for his pupils and some for others." [14]

---

[13] Dirk J. Struik, *Yankee Science in the Making* (Boston: Little Brown & Co., 1948), pp. 47, 70–71.

[14] Robert P. Multhauf, ed., "Holcomb, Fitz, and Peate; Three 19th Century American Telescope Makers" (paper 26 in *Contributions from the Museum of History and Technology*, U.S. National Museum Bulletin 228, Washington, 1962), p. 162.

Instruments of Metal

Pre-Revolutionary Immigrant Makers

According to present evidence, only a few makers of metal instruments emigrated from England to the Colonies before the beginning of the Revolutionary War. A slightly larger number emigrated after the war had ended. In almost every instance, the immigrant instrument makers settled in the major cities, which were the shipping centers of the new country. The reason is obvious: in these cities there was the greatest demand for nautical and other instruments.

One of the earliest immigrant instrument makers arrived in Boston in 1739. According to an advertisement that appeared in *The Boston Gazette* in the issue of July 16-23, 1739, there had

> Arriv'd here by Capt. *Gerry* from *London* John Dabney, junr. who serv'd his time to Mr. Jonathan Sisson, Mathematical Instrument Maker to his Royal Highness, the Prince of Wales. Makes and sells all sorts of Mathematical Instruments, in Silver, Brass, or Ivory, at Reasonable Rates, at Mr. Rowland Houghton's Shop the north side of the Town Huse in Boston.
>
> N.B. Said Dabney, sets Loadstones to a greater Perfection than any heretofore.

Dabney's master, Jonathan Sisson (1694–1749) originally of Lincolnshire, with a shop in the Strand, London, was a well-known maker of optical and mathematical instruments in the early decades of the 18th century. He was particularly noted for the exact division of scales, and examples of his work are to be found in the major collections.

Dabney's name appeared again several years later, in the Supplement to the *Boston Evening Post* for December 12, 1743, and again in the *Boston Evening Post* for December 19 of the same year, with the following advertisement:

> To be shown by John Dabney, Mathematical Instrument maker, in Milk Street, Boston, on Monday, Tuesday and Thursday Evenings, from five to eight o'clock, for the Entertainment of the Curious, the Magic Lanthorn an Optick Machine, which exhibits a great Number of wonderful and surprising Figures, prodigious large, and vivid, at Half a Crown each, Old Tenor.

In New York City, one of the earliest immigrant instrument makers was Charles Walpole, who established a shop at a corner in Wall Street, according to a notice in the May 26, 1746, issue of the *New York Evening Post.* The announcement stated that Walpole was a "citizen of London" and that at his shop "all sorts of Mathematical Instruments, whether in silver or brass, are made and mended . . . ."

In the May 21, 1753, issue of *The New York Gazette or The Weekly Post Boy* there was an announcement by the widow of Balthaser Sommer who lived on Pot-Baker's Hill in Smith Street in New York City and who advertised herself as a "grinder of all sorts of optic glasses, spying glasses, of all lengths, spectacles, reading glasses for near-sighted people or others; also spying glasses of 3 feet long which are to set on a common Walking-Cane and yet be carried as a Pocket-Book."

John Benson emigrated from Birmingham, England, and established a lapidary and optical store in May 1793 at 12 Princess Street in New York, where he produced miniatures, lockets, rings, glasses, "as well as Spectacles, single reading and burning glasses, and where he also polished scratch'd glasses." In July 1797 he moved to 106 Pearl Street where he sold green goggles, thermometers, and opera and spy glasses, in addition to an assortment of jewelry. In September 1798 he was established at a new location, 147 Pearl Street, "At the sign of The Green Spectacles" where he specialized in optical goods. He featured for rent or sale a "Portable Camera Obscura" for the use of artists in drawing landscapes. His advertisements chronicled each change in location in the issues of *The New York Daily Advertiser.*

A craftsman whose name is well known in scientific circles was Anthony Lamb, who advertised in 1753 as a mathematical instrument maker living on Hunter's Key, New York. He claimed that he could furnish

> Godfrey's newly invented quadrant, for taking the latitude or other altitudes at sea; hydrometers for trying the exact strength of spirits, large surveying instruments in a more curious manner than usual; which may be used in any weather without exception, small ditto which may be fixed on the end of a walking stick, and lengthened to a commodious height, gauging instruments as now in use, according to an act of assembly with all other mathematical instruments for sea or land, by wholesale or retail at reasonable rates.[15]

Lamb had served an apprenticeship with Henry Carter, a mathematical instrument maker in London. In July 1724 he

[15] *New York Gazette, Revived in the Weekly Post-Boy*, January 23, 1749.

became an accomplice of Jack Sheppard, a notorious burglar, and was arrested and sentenced to the gallows in 1724. As he was awaiting execution on the gallows at Tyburn, his sentence was commuted to transportation to Virginia for a period of seven years, inasmuch as this was his first offence. After he had completed his term of seven years in Virginia he moved to Philadelphia, where he opened a shop as an instrument maker and a private school for teaching technical subjects. The curriculum included surveying, navigation, and mathematics. Although his enterprises prospered, he moved to New York. There he married a Miss Ham and established himself in a respectable position. Lamb's first advertisement in New York appeared on January 23, 1749. He died on December 11, 1784, at the age of 81, and two days later he was eulogized in *The New York Packet* where he was mentioned as "a steady friend to the liberties of America."

John Lamb (1735–1800), Anthony's son, learned and practiced his father's craft for a time and worked as a partner in the firm of A. Lamb & Son. He subsequently became a wine and sugar merchant, achieved considerable wealth, married well, and was accepted by the gentry of the city. He was a firm patriot and from 1765 he was active as the leader of the Sons of Liberty. He served in several major engagements in the American Revolution and in 1783 was brevetted a brigadier-general.[16]

The immigrant instrument makers were not confined to those working in glass, however. One of the earlier immigrant craftsmen was Charles Blundy, a London watchmaker who established himself on Church Street in Charleston, South Carolina, in 1753. He notified the public that in addition to watches he sold thermometers of all sizes and types. Presumably his merchandise was imported from England.[17] He was absent from the city between 1753 and 1760 but returned and continued in business in the latter year.

Another pre-Revolution immigrant was Thomas Harland (1735–1807), a clock maker who settled in Norwich, Connecticut, in 1773. It is claimed that he sailed from England on one of the ships carrying the tea destroyed by the Boston Tea Party. Over

---

[16] Carl Bridenbaugh, *The Colonial Craftsman* (New York: New York University Press, 1950), pp. 160–161; Isaac Q. Leake, *Memoir of the Life and Times of General John Lamb* (Albany: Munsell, 1850); Silvio A. Bedini, *Ridgefield in Review* (New Haven: Walker-Rackliffe, 1958), pp. 71, 84.

[17] Alfred Coxe Prime, *The Arts and Crafts of Philadelphia, Maryland and South Carolina, 1786–1800* (The Walpole Society, 1929), p. 230.

the course of the years his business prospered to such a degree that he hired from ten to twelve apprentices at one time. Some of the leading American 18th-century clockmakers served apprenticeships with Harland. In 1802 his newspaper notice stated that he had for sale "Surveyors Compasses, with agate centre needles; chains and Protractors . . . ."[18]

A most interesting instrument that has recently come to light is a brass sundial made in Philadelphia in 1764. The dial, about 10½ inches in diameter, is signed by the maker, "Daniel Jay Philad[a]. fecit." It is dated 1764 and inscribed with the name of the person for whom it was made, "James Pemberton." In the center is "Lat. 40," which coincides with the latitude for Philadelphia. The style of the dial is very much in the English tradition of the period, indicating that Jay was probably an emigrant trained in England.

Post-Revolutionary Immigrant Makers

A large proportion of the English craftsmen who came to the American Colonies after the Revolution settled in Philadelphia. There was John Gould for instance, a mathematical instrument maker from London who had opened a shop at 47 Water Street by 1794. He sold nautical, surveying, and optical instruments as well as mirrors, presumably all imported from England. He moved to 70 South Front Street "At the Sign of the Quadrant" in 1796. He was succeeded in business in 1798 by Thomas Whitney, another emigrant from London. Whitney made and sold instruments (see fig. 85) in Gould's former shop, and featured also a vast array of department store merchandise. John Whitney, who may have been his son, was listed at the same address in the Philadelphia directory of 1801 as a "Mathematical Instrument Maker and Optician."[19]

In the Philadelphia directory and register for 1821 Thomas Whitney advertised that he

> . . . presents his sincere thanks to his friends and the public and respectfully soliciting the continuation of their favors, wishes to inform them that he has devoted his attention principally to the making of surveying com-

[18] Penrose R. Hoopes, *Connecticut Clockmakers of the Eighteenth Century* (New York: Dodd Mead & Co., 1930), p. 86; *The Norwich Courier*, February 10, 1802.

[19] Harrold E. Gillingham, "Some Early Philadelphia Instrument Makers," *The Pennsylvania Magazine of History and Biography* (1927), vol. 51, no. 3, p. 303–305.

passes for 16 years past, and has made 500 of them; the good qualities of which are well known to many surveyors, in at least 16 of the States and Territories of the Union . . . . [he also makes] many other instruments, protractors, gunner's Calibers and quadrants, etc.

George Evans was another instrument maker who arrived from London after the end of the Revolution. He established himself in a shop at 33 North Front Street in 1796, where he sold imported instruments as well as stationery, Bibles, and cloth. He died in 1798.[20]

Thomas Dring, who migrated from England, settled in Westtown Township of Chester County, Pennsylvania, where he was first noted in the tax records of 1786. He married Hanna Griffith, a native of the region, and their son, Jeptha Dring, subsequently was mentioned as a carpenter by trade, and a vagrant by inclination, who could quote Shakespeare from memory. According to local legend, Dring raised money from a number of townspeople for the purpose of purchasing clocks for them in England. He set sail for his homeland in about 1798 and never returned.

Although the tax records for 1796 described Dring as an "Optician" he was also a clockmaker and maker of scientific instruments. At least three of his tall-case clocks have survived, and a stick type of barometer which he made for Edward and Hannah Hicks in 1796. The instrument is now in the collection of the Chester County Historical Society. It measures 39 inches in height, and is signed on the thermometer dial THOMAS DRING/ West Chester. This instrument (fig. 14) is one of the very rare barometers produced in America in the 18th century.

Another craftsman who emigrated from England was Robert Clark, who opened a shop at 5½ Church Street in Charleston, South Carolina, in 1785. In that year he announced himself as a

> Math., Optical and Philosophical Instruments maker and Clockmaker from London . . . . As the Advertiser has lately had an opportunity of working and receiving instruction under the first masters in the above branches in Great Britain, flatters himself that he shall give satisfaction to those who may be pleased to favor him with their orders . . . for Surveyors compasses, Quadrants, Telescopes, Microscopes, Spirit Levels, etc.[21]

W. Fosbrook was another craftsman originally from London. He was a cutler and maker of surgical instruments, with a shop in Beekman's Slip in New York City in 1786 or earlier. He specialized in leg irons and rupture trusses, and he made instruments and

---

[20] Ibid., p. 304.
[21] *Charleston Evening Gazette*, July 24, 1785; PRIME, op. cit. (footnote 17), p. 234.

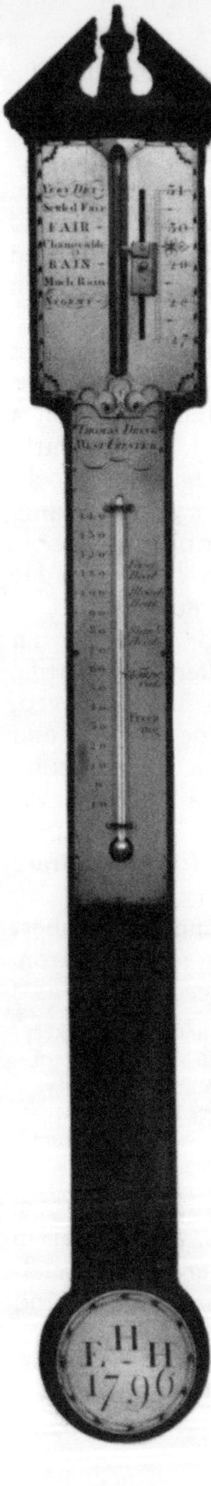

Figure 14.—Barometer made in 1796 by Thomas Dring of West Chester, Pa., for Edward and Hannah Hicks. Photo courtesy the Chester County Historical Society.

files for setting the teeth as well as standard items for surgeons.[22]

Several immigrant instrument makers established themselves in Philadelphia during the same period. John Denegan (or Donegan), stated to have been "late from Italy," moved his shop in March 1787 to the corner of Race and Fourth Streets at "the sign of the Seven Stars".[23] There he made barometers and thermometers as well as glasses for philosophical experiments. It seems too much of a coincidence that in October 1787 an instrument maker named Joseph Donegany established a shop at 54 Smith Street in New York City,[24] where—according to an advertisement in the October 17, 1787, issue of *The New York Daily Advertiser*—he made "thermometers, barometers and sold hydrostatic Bubbles and hygrometers for proving spirits, and also . . . glasses for experimental purposes." It is probable that Denegan and Donegany were one and the same; since Denegan was stated to have been of Italian origin, the name may originally have been "De Negani."

Joseph Gatty advertised himself as an "Artist from Italy" with a shop at 341 Pearl Street in New York City where he "made and sold every simple and compound form of barometer and thermometer as well as curious Hygrometers for assaying spirits which show the actual strength with the greatest precision and are not liable to be corroded, in addition to several new Philosophical Instruments of his own invention, and all types of artificial fireworks."[25] By 1796 Gatty (or Gatti?) had moved to Philadelphia where he had a shop at 79 South Front Street and advertised the same items that had appeared in his advertisements in New York. The Philadelphia directory for 1800 listed Gatty as a "Weather Glass Maker."[26]

Native American Makers

Comparatively speaking, the greater proportion of the early American instrument makers were native born. Among these were to be found a substantial number of artisans trained as clockmakers who subsequently produced scientific instruments to meet the surveying and nautical needs of their communities.

[22] RITA S. GOTTESMAN, *The Arts and Crafts in New York, 1777–1799* (New York: New York Historical Society, 1954), pp. 220–221.

[23] *The Pennsylvania Evening Herald*, March 17, 1787.

[24] GOTTESMAN, op cit. (footnote 22), pp. 311–312.

[25] *The Diary, or Evening Register*, November 3, 1794.

[26] GILLINGHAM, op. cit. (footnote 26), p. 306.

Figure 15.—James Wilson, first American maker of globes. From a sketch by John Ross Dix in *Ballou's Pictorial Drawing Room Companion* (Boston, 1857), vol. 12, p. 156.

Together with the other craftsmen throughout the colonies who established and advertised themselves specifically as instrument makers, they produced a large number of the mathematical instruments used in the American Colonies in the 18th century. A careful study of their regional distribution reveals that most of them were concentrated in the major coastal cities of commerce.

New Hampshire

Among the artisans who combined clockmaking with instrument making before the beginning of the 19th century was Benjamin C. Gilman (1763–1835) of Exeter, New Hampshire. He made mathematical instruments and clocks in addition to working as a silversmith, clockmaker, and hydraulic engineer.

Vermont

A New England instrument maker who had a most unusual career was James Wilson (1763–1855) of Bradford, Vermont. He was a native of Francestown, New Hampshire, where he was born in a log cabin and brought up on a farm. In 1796 he purchased his own farm, at Bradford.

Figure 16.—Globe made by James Wilson (1763–1855) of Bradford, Vermont. Diameter is 13 in. Photo courtesy Houghton Library, Harvard University.

When a young man of 36 he saw a pair of globes at Dartmouth College in neighboring Hanover and tried to duplicate them. He made balls of wood turned from solid blocks, covered them with paper, and finished them off with lines and drawings. He later improved this method by coating the wooden balls thickly with layers of paper pasted together. He then cut the globes into hemispheres, removed the wooden molds, and joined the paper shells to make the globes.

Wilson next proceeded to procure copper plates of the necessary sizes for his globes, and he projected his maps on them in sections. He received a few lessons in engraving from Amos Doolittle of New Haven, but he was otherwise completely self-taught.

Wilson exhibited the first edition of his globes in Boston in 1814. They created a sensation, and many persons asked to see the maker, but Wilson was reluctant to come forward because of his coarse clothing and rustic manners. He was greatly encouraged, however, by the public interest in his work, and he continued to make globes in Bradford (see fig. 16). In about 1815 Wilson and his three sons, all of whom were as ingenious as the father, formed a company to manufacture globes in Albany. There they produced terrestrial and celestial globes, the latter showing as many as 5,000 stars. Wilson produced a new set of plates in 1826 and made globes in several sizes. Even after he had reached the age of 83 years he constructed an excellent planetarium, engraving the large copperplate himself.

Wilson was married three times and was the father of 14 children. He died at the age of 92 in March 1855 at Bradford.[27]

Massachusetts

A surprisingly small number of the Massachusetts craftsmen working before the end of the 18th century produced scientific instruments. Among the very earliest were several members of the King family of Salem. Daniel King (1704–1790) was born in Salem on November 17, 1704. At the time of his death Rev. William Bentley spoke of him as a "maker of Mathematical Instruments" and a "teacher of Mathematics."[28]

---

[27] Edwin Valentine Mitchell, *The Romance of New England Antiques* (New York: A. A. Wyn, 1950), pp. 257–160; Kimball op. cit. (footnote 7).

[28] William Bentley, *Diary of William Bentley, D. D.* (Salem, Mass.: 1905), vol. 1, p. 182, vol. 2, p. 414.

Figure 17.—Brass surveying compass made by Stephen Greenleaf (fl. 1745) of Boston. Photo courtesy New Hampshire Historical Society, Concord.

Following Daniel's death, his business in instruments was inherited by his son Benjamin King (1740–1804), of Salem. Benjamin specialized in producing nautical instruments, and several of his Davis quadrants have survived in public collections. When he died on December 26, 1804, Reverend Bentley wrote that King was ". . . a Mathematical Instrument maker, in that branch which immediately regarded practical navigation by quadrant and compass. He supported a very good character through life & was much esteemed."[29]

Another of the very early mathematical instrument makers in Massachusetts was Stephen Greenleaf (see fig. 17), who kept a shop on Queen Street opposite the prison in Boston where

> He makes and Mends all Sorts of Mathematical Instruments, as Theodolites, Spirit Levels, Semicircles, Circumferences, and Protractors, Horizontal

[29] Ibid., vol. 3, p. 130.

and Equinoctial Sun Dials, Azimuth and Amplitude Compasses, Eliptical and Triangular Compasses, and all sorts of Common Compasses . . . N.B. He sets Load Stones on Silver or Brass, after the best manner.[30]

Jonathan Dakin worked as a mathematical balance-maker "at the Sign of the Hand & Beam, opposite to Dr. Colman's Meeting House" where he made a variety of scale beams in 1745.[31]

An interesting advertisement by Rowland Houghton appeared in the January 17–24, 1737, issue of the *Boston Gazette*. Houghton announced that he had "lately improv'd on his new Theodolite, by which the Art of Surveying is rendered more plain & easy than heretofore." Houghton was active in the political scene in Boston, as evidenced by the fact that in various issues of *The Boston Gazette* for January and February 1739 he is listed variously as "Commissioner," "Proprietors' Clerk" and as "Collector."

Isaac Greenwood, Jr. (1730–1803), was born at Cambridge, where he married Mary I'ams in 1757. He maintained a shop where he combined the business of mathematical instrument maker and ivory turner, and also imported hardware. After the Revolution, he engaged in dentistry, specializing in making artificial teeth and in the manufacture of "umbrilloes." Paul Revere apparently did printing for him on five different occasions between 1762 and 1774, and in about 1771 engraved his trade card, which read:

> ISAAC GREENWOOD, Ivory Turner Next door to Doctor John Clark's at the North End Boston. Turns all sorts of work in Ivory, Silver, Brass, Iron, Horn, Wood, etc. Such as Billiard Balls, Tea Boards, Scallop[d] and Plain Salvers, Decanters . . . .[32]

Isaac's father, Isaac Greenwood, Sr., was "Professor of Mathematicks and Natural and Experimental Philosophy" at Harvard. In the *Boston Gazette* for February 19–26, 1728, there appeared the following notice of his installation:

> On the 13th of this Month at Ten in the Morning, The Honorable & Reverend Overseers of the College at Cambridge, met the Corporation in the College Hall, to Inaugurate Mr. Isaac Greenwood into the Office of Professor of the Mathematicks, and Natural and Experimental Philosophy, lately founded by that great and living Benefactor to this Society, Mr. Thomas Hollis of London Merchant. The Rev. President being detain'd by illness,

---

[30] *Boston Gazette*, June 18, 1745.

[31] Ibid., November 12, 1745.

[32] Clarence S. Brigham, *Paul Revere's Engravings* (Worcester, Mass.: American Antiquarian Society, 1954), p. 118; Bernard W. Wienberger, *Introduction to the History of Dentistry* (St. Louis, Mosby Co., 1948), 2 vols., vol. 2, pp. 119–134; Isaac J. Greenwood, *The Greenwood Family*, 1934, pp. 68–78.

Mr. Flint the Senior Fellow perform'd the part of Moderator, began with Prayer, and then Pronounc'd a Latin Oration proper to the Occasion: Mr. Wiggleworth Divinity Professor, read the Founders Instructions. Mr. Greenwood took the Oaths and made the Declarations required in them: and pronounc'd a Latin Oration. The Rev. Mr. Appleton Pray'd: and Singing part of the 104 Psalm concluded the Solemnity. After which the Overseers & Corporation repair'd to the Library; till the Publick Dinner in the Hall was ready, where all the Gentlemen Spectators of the Solemnity were hansomely Entertained.

Greenwood continued to teach privately for a decade. In various issues of *The Boston Gazette* of 1738 and 1739 he featured an advertisement, the text of which always stated:

Such as are desirous of learning any Part of Practical or Theoretical Mathematics may be taught by Isaac Greenwood, A.M. &c. in Clark's Square, near the North Meeting House, where Attendance will be given between the Hours of 9 and 12 in the Forenoon, and 2 and 5 in the Afternoons.

N.B. Instructions may also be had in any Branch of Natural Philosophy, when there is a sufficient Number to attend.[33]

John Bailey II (1752–1823) of Hanover and Lynn, Massachusetts worked as a clockmaker from about 1770. His father, John Bailey I, and his brothers Calvin and Lebbeus also were clockmakers. Bailey married Mary Hall of Berwick, Maine, and settled in Hanover where he made scientific instruments and clocks. A brass surveying compass in the collection of the New York Historical Society is inscribed "J. BAILEY HANOVER 1804."[34]

Undoubtedly the best known instrument maker in Massachusetts was Joseph Pope (1750–1826), of Boston, who was described by contemporaries as the "local mathematician, watch-maker and mechanical genius." In 1787 he completed the construction of a gear-driven orrery displaying the motions of the solar system in a horizontal plane with eccentric and inclined orbits. At each of the twelve corners were mounted cast bronze figures, claimed to have been carved in wood by Simeon Skillin and cast in bronze by Paul Revere. Although the instrument was made for Harvard, the university lacked funds for its purchase. Accordingly, it held a public lottery which realized a substantial sum in excess of the £450.3.0 paid to Pope, and the orrery was delivered in De-

[33] *Boston Gazette*, November 6–13 and November 20–27, 1738, March 26–April 2 and April 2–9, 1739.

[34] Brooks Palmer, *The Book of American Clocks* (New York: Macmillan Co., 1950), pp. 141–142.

cember 1788.[35] The orrery (fig. 18) has survived and is part of the collection of historical scientific instruments at Harvard University.

According to a statement in the *Boston Gazette* for February 16, 1789, an apparatus for displaying planets in their proper orbits by means of wires was made and exhibited in Boston by Bartholomew Burges.

Mention must also be made of several members of the Folger family of Nantucket, Massachusetts. Peter Folger (1617–1690), founder of the American branch of the family, emigrated from Norfolk, England, in 1635 and occupied himself in Nantucket as blacksmith, schoolmaster, watchmaker, and surveyor. He was a grandfather of Benjamin Franklin. Another notable descendant was Maria Mitchell (1818–1889), professor of astronomy and director of the observatory at Vassar College.

The best known member of the family was Walter Folger, Jr. (1765–1849), a self-taught clockmaker and watchmaker with great interest in the sciences. A telescope that he produced about 1818 was considered to be the finest in the country at that time. His greatest achievement was a tall case astronomical clock that he devised and constructed; it was completed in 1790 and is considered to be the most complicated domestic clock on record.[36] Folger also produced quadrants and compasses, and made astronomical observations. His observations of the solar eclipse of September 17, 1811, were published in 1815 in *Memoirs of the Academy of Arts and Sciences*.

Probably one of the most significant of the surviving early American scientific instruments is a pair of gunners' calipers made and used by PAUL REVERE (1735–1818) of Boston. The calipers are made of incised brass, measuring 7 inches in length and 1¾ inches in width. They are signed on the reverse side with the name "Revere" in the style of script signature used by this maker in many of his engravings. The design of the instrument is substantially different from that which is commonly found in English, French, and German gunners' calipers of the period, and was probably Revere's own. (See figs. 19, 20.)

---

[35] *Massachusetts Magazine* (1789), vol. 1, pp. 36, 37; *Boston Gazette*, January 12, 1789; I. BERNARD COHEN, *Some Early Tools of American Science*, (Cambridge: Harvard University Press, 1950), pp. 64–65, 157; HARROLD E. GILLINGHAM, "The First Orreries In America," *Journal of the Franklin Institute* (1940), vol. 229, pp. 92–97.

[36] WILL GARDNER, *The Clock that Talks and What It Tells* (Nantucket Whaling Museum, 1954), pp. 34–40, 97, 106.

Figure 18.—Orrery by Joseph Pope completed in 1787 for Harvard University. Engraved plates and bronze figures were made by Paul Revere. The orrery is 6½ ft. in diameter and 6½ ft. high. The twelve figures at the corners are said to have been carved in wood by Simeon Skillin and cast in bronze by Paul Revere. Photo courtesy Harvard University.

It is believed that these calipers, which are preserved in the collection of the Bostonian Society in Boston, were probably used by Revere in 1775–1776. This was the period during which he was in charge of ordnance repairs for the Continental Army, and involved in various ventures for the manufacture of gunpowder and the casting of cannon. There is no evidence of other scientific instruments made by Revere, lending some weight to the belief that these calipers were made for his own use.

Other Massachusetts instrument makers include Gideon Fairman (1774–1827) of Newburyport who was a partner of William Hooker in the firm of Hooker & Fairman, which dealt in mathematical instruments before 1810.[37] Fairman later moved to Philadelphia,

[37] PALMER, op. cit. (footnote 34), p. 190.

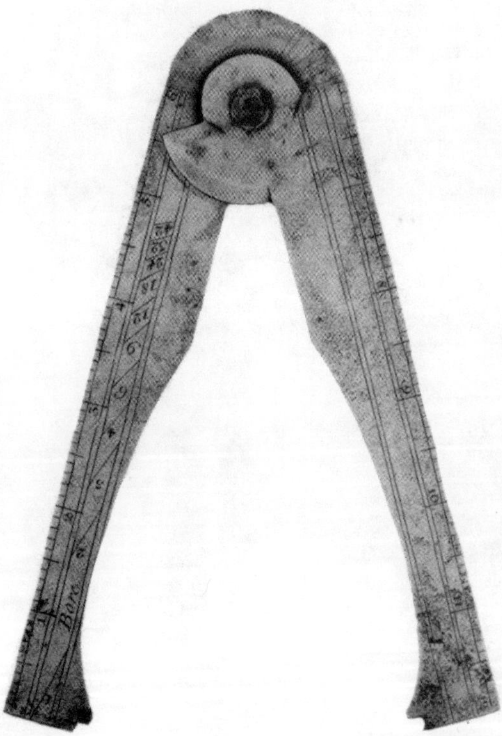

Figure 19.—Brass gunnery calipers made and probably used by Paul Revere (1735–1818). The calipers are 7 in. long and 1¾ in. wide.

Figure 20.—Reverse side of gunnery calipers, showing the inscribed signature. Photos courtesy the Bostonian Society, Boston, Mass.

where he was associated with the engraving firm of Draper, Murray & Fairman.

At the end of the 18th century Samuel Emery was making mathematical instruments in Salem, at the same time that John Jayne was engaged in the same work in that community.[38]

John Potter of Brookfield, Massachusetts, produced surveying instruments in the last quarter of the 18th century. A graphometer signed with his name and dated 1785 is in the collection of the firm of W. & L. E. Gurley in Troy, New York.

Rhode Island

One of the earliest and most important of the instrument makers of Rhode Island was Benjamin King (1707–1786), of Newport. He was the son of Capt. Samuel King of Salem, Massachusetts, where he was born and baptized on March 13, 1707. He was a brother of Daniel King of Salem. Benjamin eventually moved to Newport, where he married Mary Hagger in July 1742. They had four children: Benjamin, Mehitable, Samuel, and Mary. He established himself as a respectable businessman in the community, and in 1759 or 1760 he became the senior partner in the importing and retailing firm of King & Hagger, "near the sign of Mr. Pitt," dealing in general merchandise, mathematical and nautical instruments, and stationery. William Hagger was probably the junior partner, and may have been King's brother-in-law. King began making his own instruments for sale, surviving examples dated as early as 1762. The partnership was dissolved early in the 1760's. In 1766 Benjamin King was importing, making, and selling quadrants and other instruments "At the Sign of the Mathematical Instruments" next to the Golden Eagle on Thames Street. His son Samuel King occupied the same premises, where he dealt in paints and artists supplies.

When the British occupied Newport, King moved to North Kingstown, but he returned after the British vacated the city. He was 79 when he died in 1786, and his son Samuel King succeeded him in business.[39]

William Guyse Hagger (c.1744–1830?), born in Newport, is believed to have been the son of William Hagar and Mary Knowlton. He was a quadrant maker (see fig. 21). In 1774 he headed a house-

[38] Joseph B. Felt, *Annals of Salem* (Salem, Mass.: Ives, 1827), vol. 2, p. 173.

[39] Howard M. Chapin, "Davis Quadrants," *Antiques* (November 1927), vol. 12, no. 5, pp. 397–399; also Rufus King, *Pedigree of King of Lynn* (Salem, Mass., 1891).

Figure 21.—Davis quadrant or backstaff made and signed by William Guyse Hagger of Newport, Rhode Island, about 1760–1770. USNM 319029.

hold that consisted of his wife, five children, and a colored servant. Whether it was he or his father who was the partner of Benjamin King cannot be determined with certainty. When Newport was occupied by the British, Hagger moved to Cranston, where he joined the Pawtuxet Rangers and served as a sentinel at Pawtuxet Fort in 1778. No members of the Hagger family appear in the 1790 census of Newport, but a William Hager is reported as having died in Boston in 1830 at the age of 82. It seems likely from the age and dates that it was William Hagger the elder who worked as a partner in the firm of King & Hagger, which was established in 1759 or 1760.[40]

Another instrument maker of Rhode Island was William Hamlin (1772–1869). He had established himself in Providence by the beginning of the 19th century in the manufacturing and repairing of mathematical and nautical instruments, for which there was an

[40] Chapin, op. cit. (footnote 39), pp. 398–399.

active market in that city. Hamlin was one of the first engravers in America and the first in Rhode Island. He designed and engraved banknotes for many banks in the State and for other institutions. At the same time he carried on a general trade in the sale of musical instruments. Hamlin moved his shop several times, but from 1847 until his death he worked at "The Sign of the Quadrant" (see fig. 22) at 131 South Water Street. He was equally interested in optics and astronomy, and it has been claimed that he constructed the first telescope in America. It is well established that he worked for many years to perfect a reflecting telescope for his own use.[41]

Instruments were made also by Paul Pease, who may have been the husband of the daughter of Nathaniel Folger of Nantucket. This Elizabeth Folger Pease, wife of a Paul Pease, was born in 1720 and died in 1795. Little is known about Pease except for the name "Paul Pease 1750" inscribed on a quadrant in the collection of the Rhode Island Historical Society.[42]

Connecticut

The clockmakers who worked in Connecticut during the span of the 18th century numbered almost a hundred. Yet only a half dozen appear on record to have made or sold instruments in addition to clocks Among these were several members of the Doolittle family, including Isaac Doolittle (1721–1800) of New Haven. In 1763 he advertised that he sold surveying compasses in addition to clocks, watches, bar iron, and chocolate.[43] His son Isaac Doolittle, Jr. (1759–1821), also of New Haven, established a shop of his own, which he advertised in 1781 as having

> Compasses of all kinds, both for sea and land, surveyors scales, and protractors, gauging rods, walking sticks, silver and plated buttons, turned upon horn; also clocks and watches made and repaired . . . .[44]

Although not very active as a clockmaker, Isaac Jr. appears to have specialized more in the production of surveying and nautical instruments. He took over his father's business just before the latter's death, and in 1799 he advertised[45]:

[41] Gladys R. Lane, "Rhode Island's Earliest Engraver," *Antiques* (March 1925), pp. 133–137.

[42] Chapin, op. cit. (footnote 39), p. 399.

[43] Hoopes, op. cit. (footnote 18), pp. 70–72.

[44] *The Connecticut Journal*, June 7, 1781.

[45] Ibid., May 22, 1799.

Figure 22.—Trade cards of William Hamlin (1772–1869), engraver and instrument maker of Providence, Rhode Island. In collection of Rhode Island Historical Society, Providence.

The subscriber having commenced business at the shop lately occupied by Mr. Isaac Doolittle, in Chapel Street, where he repairs watches, makes and repairs Surveyors Compasses and Chains, Brass Amplitude, plain brass and common Ship's Compasses, Gauging Rods, Quadrants, repair'd &c. every favor gratefully received by the public's humble servant, Isaac Doolittle, jun.

Enos Doolittle (1751–1806), a nephew of Isaac Doolittle, Sr., made, sold, cleaned, and repaired clocks and surveying and marine compasses from 1772 through 1788 at his shop in Hartford. He also sold these items through agents in Saybrook and Middleton.[46]

One of the best known of the Connecticut clockmakers was Peregrine White (1747–1834), of Woodstock. White was a descendant of the first Pilgrim child, and a native of Boston. After serving an apprenticeship, he worked as a clockmaker and silversmith in Boston. He was accused of forging silver spoons and left the city to settle in Woodstock. He established his own shop west of Muddy Brook Village.[47] In addition to fine tall-case clocks, for which he was noted, White also produced surveying compasses, one of which is in the collection of the U.S. National Museum (fig. 23). A similar specimen in Old Sturbridge Village is reputed to have been used for surveying the town of Southbridge, Mass.

Benjamin Hanks (1755–1824), of Mansfield and Litchfield, inserted a notice in a newspaper in 1808 to notify the public that he and his son Truman Hanks, in partnership, had "surveyors compasses upon the Rittenhouse improved plan" in addition to such other commodities as brass cannon, bells from their own foundry, clocks, goldsmith's items, and stocking looms.[48]

Ziba Blakslee (1768–1834), of Newton, worked as a clockmaker, goldsmith, and bell founder and he advertised that he made and sold surveying instruments.[49]

In New Haven, Clark Sibley and Simeon Marble organized the firm of Sibley & Marble and advertised that in addition to repairing swords and cutlasses, clocks and watches, they also repaired mathematical and surgical instruments.[50]

---

[46] *The Connecticut Courant*, December 15, 1772, and October 22, 1787; HOOPES, op. cit. (footnote 18,) pp. 66–70.

[47] HOOPES, op. cit. (footnote 18), p. 122.

[48] Ibid., pp. 79–83.

[49] PALMER, op. cit. (footnote 34), p. 159.

[50] PENROSE R. HOOPES, *Early Clockmaking in Connecticut* (New Haven: Yale University Press, 1934), pp. 8–9.

Figure 23.—Brass surveying compass made about 1790 by Peregrine White (1747–1834) of Woodstock, Connecticut. USNM 388993.

One of the instrument makers of New England who has remained relatively unknown was Benjamin Platt (1757–1833), who was born in Danbury, Connecticut, on January 3, 1757.[51] He married Adah Fairchild of the same city in 1776, and it is believed that he must have completed his apprenticeship by that date inasmuch as apprentices usually were not allowed to marry.

It is not known how long Platt worked in his native city, but by 1780 he had moved to Litchfield, where he worked in gold, silver, and brass. He became established as a clockmaker and produced tall case clocks and other types. In 1787 he was in New Milford, a town adjacent to Danbury, where he produced surveying compasses (see fig. 24). Three years later, in 1790, he was at Milford, where he invented a "Compass for measuring distance in hilly country." In 1793 he returned to New Milford, where he made a clock to order for Eli Todd, and by 1800 he had moved to Lanesboro, Massachusetts.

Ohio

Benjamin Platt was the migratory type. In 1817 he migrated from Lanesboro to Columbus, Ohio. His son, Augustus Platt (1793–1886), also made mathematical instruments (see fig. 25) in Columbus. In 1809 a grandson, named William Augustus Platt was born. When the child's mother died, Benjamin and Adah Platt adopted the boy, and when he came of age he went into the watchmaking trade. William Platt married Fanny Hayes, sister of President Hayes.[52] His shop was listed in the 1843 city directory; it was the first jewelry and clock and watch store in the community.

An interesting account of instrument making in Ohio is found in the report of a missionary, John Heckewelder. He mentioned the brothers Joseph and Francois Devacht who worked as watchmakers and instrument makers in Gallipolis, Ohio. Writing in 1792, Heckewelder stated that "the most interesting shops of the Workmen [in Gallipolis] were those of the Goldsmiths and Watchmakers. They showed us work on watches, compasses, sundials finer than I have ever beheld."

---

[51] William McCabe, "Benjamin Platt of New Fairfield, Connecticut," *Timepieces Quarterly* (November 1948), vol. 1, no. 1, pp. 26–28.

[52] Ibid.

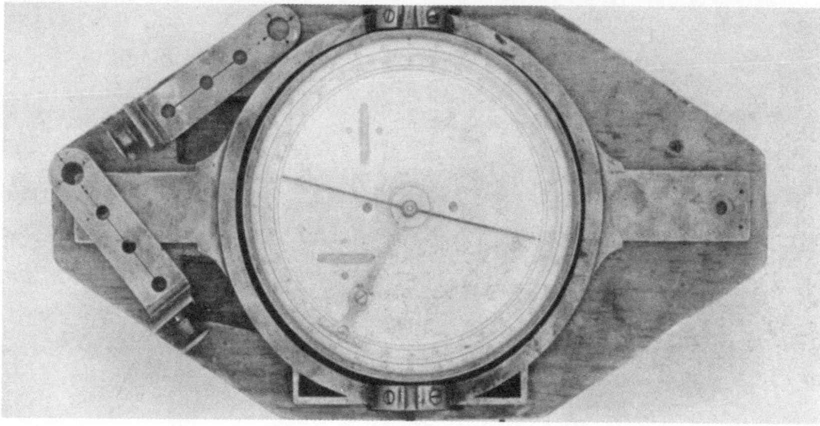

Figure 24.—Brass surveying compass made by Benjamin Platt (1757–1833) of New Milford, Connecticut, about 1795–1800. Shown in original wooden case and separately (opposite page). Photos courtesy Ohio State Museum.

New York

There were relatively few makers of mathematical instruments in New York City before the end of the 18th century. Perhaps the earliest was John Bailey, who moved from Fredericksburg, Virginia, to Fishkill, New York, in 1778. He was a cutler by trade, and he made and sold surgical instruments.[53]

"Bulmain & Dennies" at 59 Water Street in New York were the appointed agents to sell the "Perpetual Log or Distance Clock to find a ship's way at sea." The device had been patented in the United States, and one of the instruments was displayed at the bar of the Tontine Coffee House, according to an advertisement in the July 23, 1799, issue of the *New York Gazette and General Advertiser.*

H. Caritat, at 153 Broadway in New York, imported and sold "The Planispherical Planetarium." This item was described in an advertisement[54] as "a graphic representation of the earth, in twelve particular positions during its revolutionary course around the sun, as also of the Moon's revolution around the earth, together with literal description of parts and motions, etc." The advertisement also stated that Caritat sold "Carey's newly improved Terrestrial

[53] *New York Packet,* May 14, 1778.

[54] Gottesman, op. cit. (footnote 22), p. 270.

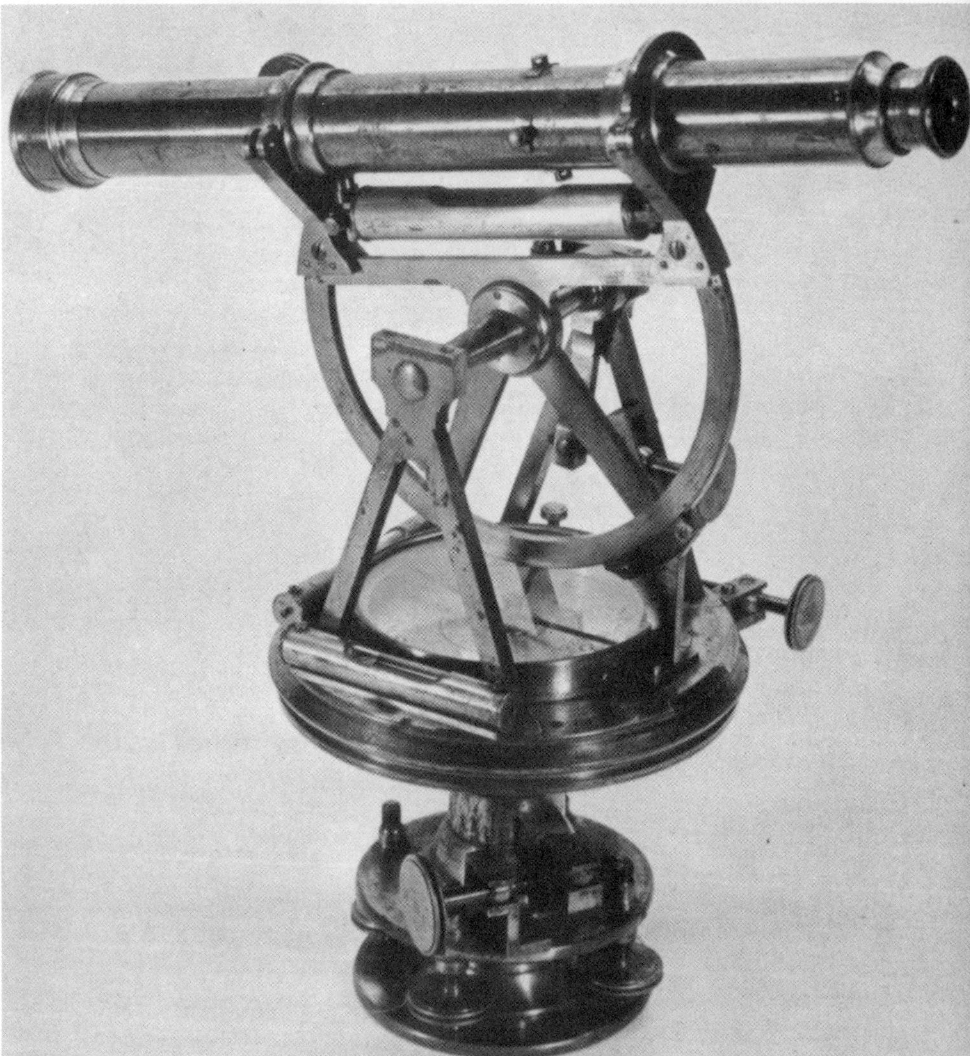

Figure 25.—Surveying theodolite made by Augustus Platt (1793–1886) of Columbus, Ohio, in the early 19th century. Photo courtesy Ohio State Museum.

and Celestial Globes which omitted the Constellary Configurations."

In 1785 M. Morris of New York City made and sold his own invention of a "Nautical Protractor for the price of One Dollar." In an advertisement in *The Independent Journal or the General Advertiser* of May 25, 1785, he explained that the device was for use in the construction of globular maps and Mercator charts. He also made another protractor for attaching to the end of a ruler for measuring distances on charts. He planned to publish a treatise on the subject of his inventions.

James Youle, a cutler and mechanician with a shop located first on Fly Street and then at 64 Water Street "at the Sign of the Cross-Knives and Gun," sold a large variety of cutlery and hardware for gun repair. He also made surgical instruments. He died in February 1786 at the age of 46 as the result of an injury to his chest from a breaking grindstone while working in his shop. He was survived by a widow and nine children and was succeeded in business by his son John Youle.[55]

New Jersey

One of the few instrument makers known to have worked in New Jersey was Aaron Miller of "Elizabeth-town." He was first noted in the New York newspapers in 1748 when he notified the public that, in addition to clocks, he made compasses, chains for surveyors, and church bells, for which he maintained his own foundry. When he died in 1771 he left all his tools to a son-in-law, Isaac Brokaw.[56]

Another craftsman who is entitled to being included as an instrument maker was Richard Wistar. When Casper Wistar died in 1752, his son Richard succeeded him as owner of the famous glass works. In addition to window glass and glassware, Richard Wistar also produced such special products as retorts for use in chemistry and "electerizing globes and tubes," as well as bottles for Leyden jars that Benjamin Franklin had urged him to attempt in the early 1750's.[57]

---

[55] *New York Packet*, February 3, 1785, and February 27, 1786, and *New York Daily Advertiser*, February 8, 1787.

[56] *The New York Gazette Revived in The Weekly Post-Boy*, January 4, 1748.

[57] Bridenbaugh op. cit. (footnote 16), p. 63; Frederick W. Hunter, *Stiegel Glass* (Boston: Houghton Mifflin, 1914), pp. 157–161.

Delaware

George Crow (ca. 1725–1771/72) of Wilmington, Delaware, was apparently well established as a clockmaker in the community by the time of his marriage in 1746 to Mary Laudonet. They had four children, and Crow's two sons followed his trade. George Crow was active in civic affairs, and in addition to clocks, he produced surveying compasses, several of which have survived.[58]

Maryland and Virginia

Brief mention has already been made of the Chandlee family of clockmakers and instrument makers of the 18th century. The founder of the line and first of interest was Benjamin Chandlee, Sr., who migrated in 1702 from Ireland to Philadelphia, where he was apprenticed to Abel Cottey, clockmaker, and eventually married his daughter. His son Benjamin Chandlee, Jr. (1723–1791), worked as a clockmaker in Nottingham, Maryland, where he produced instruments as well as clocks. A fine example of a brass surveying compass—inscribed with his name, and which is believed to have been made for the Gilpin family in about 1761—is on exhibition in the Chester County Historical Society. He had four sons, and a few years before his death he established the firm of Chandlee & Sons, the name of which was changed to Ellis Chandlee & Brothers a year before he died.

The oldest of Benjamin Jr.'s four sons was Goldsmith Chandlee (c.1746–1821). After serving an apprenticeship with his father, Goldsmith moved to Virginia and worked near Stephensburg (now Stephens City). He eventually established himself at Winchester and built a brass foundry and a shop where he produced clocks, surveying compasses, sundials, apothecary and money scales, surgical instruments, compasses, telescopes, and other items in metal. Numerous examples of his clocks and instruments have survived. Their fine quality attests to the claim that he was one of the foremost craftsmen of the 18th century. Several of his surveying compasses exist in modern collections. An instrument (fig. 26) that he made about 1794 for a surveyor named Robert Lyle is in the writer's collection; an almost identical instrument that Chandlee made for Lawrence Augustine Washington, George Washington's nephew, is exhibited in the library at Mount Vernon, Virginia.

---

[58] Henry C. Conrad, "Old Delaware Clockmakers," *The Historical and Biographical Papers of the Historical Society of Delaware* (1897), vol. 3, chap. 20, pp. 4–34.

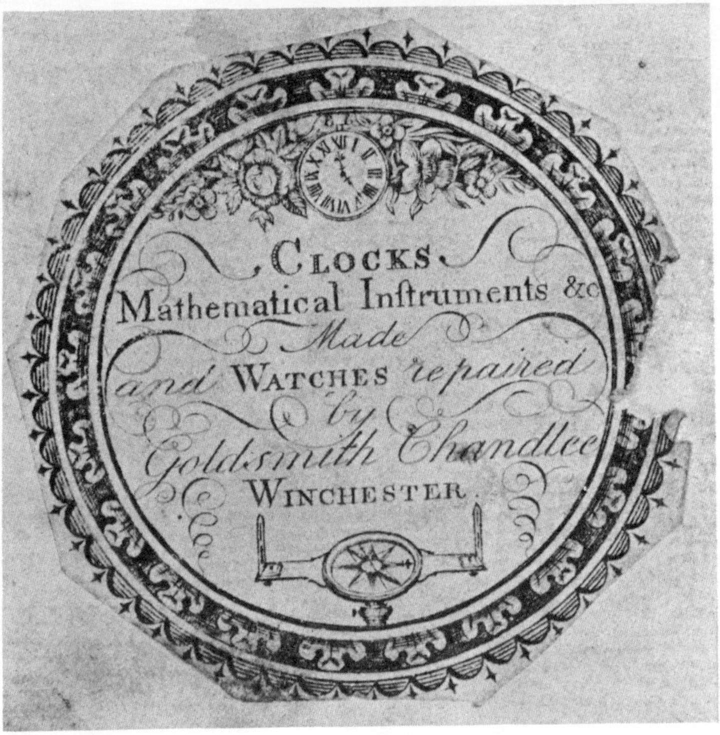

Figure 26.—The label of Goldsmith Chandlee. In the collection of Ohio Historical Society, Ohio State Museum.

Ellis Chandlee (1755–1816) also was apprenticed to his father, and he worked with his brothers in the shop. He established the firm of Ellis Chandlee & Brothers, in 1790, shortly before his father's death. The firm was dissolved in 1797 when the youngest brother, John Chandlee, left the firm. Ellis continued in partnership with his other brother, Isaac Chandlee (1760–1813), until about 1804, producing clocks, surveying instruments, and other metal articles. Their products were signed "Ellis and Isaac Chandlee, Nottingham," or, in the case of a surveying compass in the collection of the Chester County Historical Society, "E. & I. Chandlee, Nottingham." Isaac Chandlee also produced clocks and instruments under his own name only, for there are a number of surviving clocks and surveying compasses signed in such manner (see fig. 28).[59]

[59] EDWARD E. CHANDLEE, *Six Quaker Clockmakers* (Philadelphia: Historical Society of Pennsylvania, 1943), pp. 70, 193, 212, 220–223.

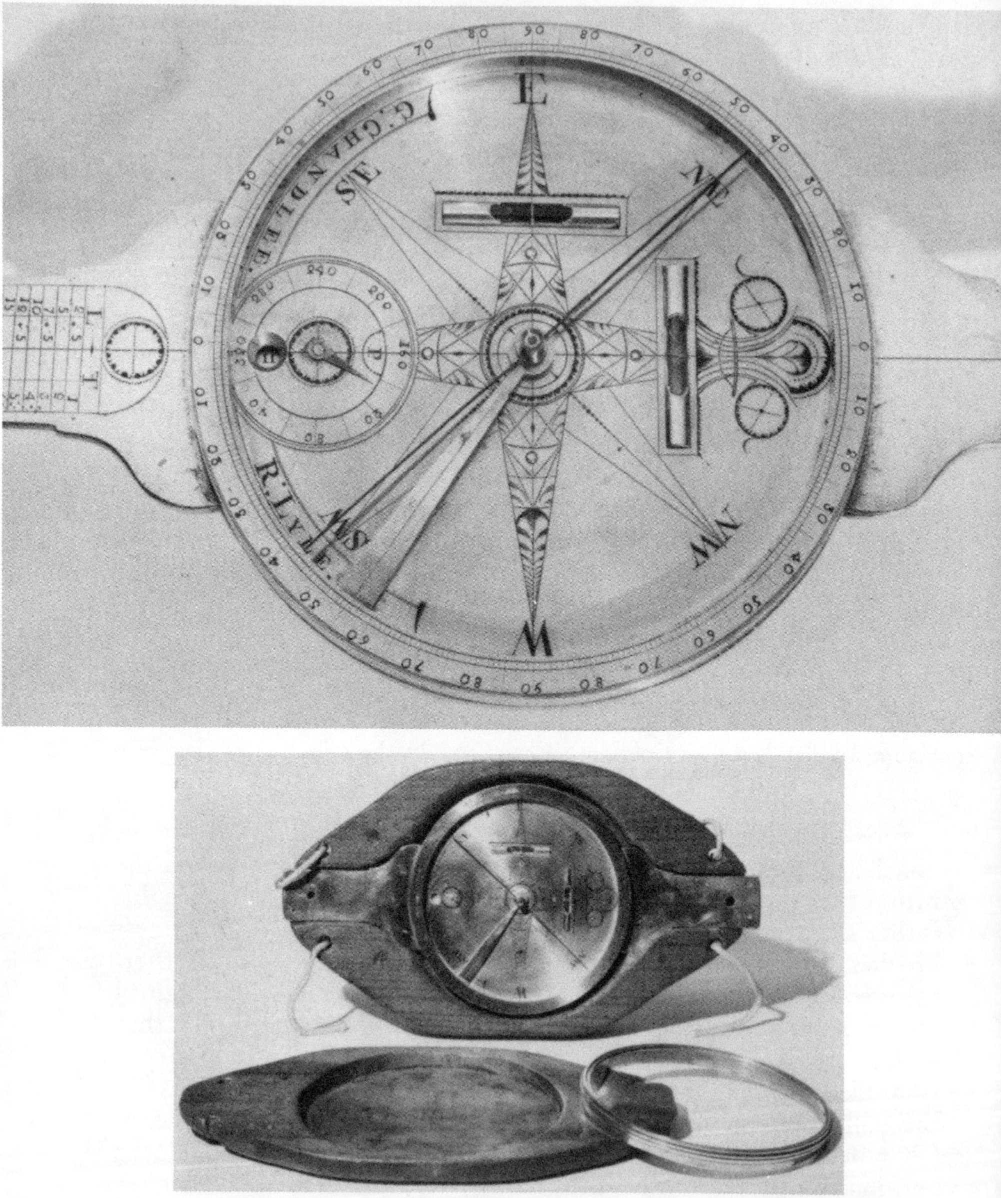

Figure 27.—Brass surveying compass with outkeeper made by Goldsmith Chandlee (c. 1746–1821) of Winchester, Virginia, for Robert Lyle. Over-all length, 14½ in.; diameter, 7 in. Instrument, in original wooden case, bears ink signature of Robert Lyle. In collection of the writer.

One of the most important craftsmen of Maryland was Frederick A. Heisely (1759–1839). A native of Lancaster, Pennsylvania, he served an apprenticeship there with John Hoff, the master clockmaker, from 1777 to 1783. Heisely served in the Revolution. In 1783, presumably upon the completion of his apprenticeship, he married Catherine Hoff, the clockmaker's daughter. He moved to Frederick, Maryland, where he established his own clockmaking shop and where he specialized in making mathematical instruments. A tower clock made in Frederick is in the collection of the U.S. National Museum. Heisely returned to Lancaster to become Hoff's partner, and worked with him until 1802. He then moved his shop to Harrisburg and worked there until 1820. He moved once more, this time to Pittsburgh where he advertised himself as a "Clock, Watch and Instrument Maker," with a shop at No. 6 St. Clair Street.

George Heisely (1789–1880), Frederick's son, who was born at Frederick, Maryland, achieved note in his own right as a maker of clocks and instruments. He worked at Second and Walnut

Figure 28.—Brass surveying compass made by Goldsmith Chandlee for Laurence Augustine Washington in about 1795. In the library at Mount Vernon. Photo courtesy the Mount Vernon Ladies Association of the Union.

Streets in Harrisburg. He is credited with being the person who selected the melody of "To Anacreon in Heaven" for "The Star-Spangled Banner," while he was serving as a member of the Pennsylvania State Militia.[60]

Pennsylvania

A number of instrument makers worked in Philadelphia, which was one of the important shipping centers during the 18th century and consequently one of the important markets for nautical instruments.

Probably the earliest Philadelphia instrument maker of record was Thomas Godfrey (1704–1749) who was born in Bristol Township. After serving an apprenticeship, Godfrey developed his own business as a glazier and plumber. He is stated to have done the major part of the glazing of the State House in 1732, as well as similar work on Christ Church. He also worked for Andrew Hamilton and for James Logan.

Godfrey had a natural inclination and interest in science and mathematics, which may have been further encouraged by his friendship with Benjamin Franklin, who resided in the same house. Godfrey was also a fellow member of Franklin's Junto.

In 1730 Godfrey invented an improved backstaff or Davis quadrant, and loaned the instrument to Joshua Fisher to be used in the latter's survey of Delaware Bay. It is claimed that the location of Cape Henlopen was established on Fisher's map (published in London in 1756) by means of Godfrey's instrument. James Logan became interested in the improved backstaff invented by Godfrey and at Logan's request, the instrument was taken on a voyage to the West Indies by a Captain Wright for the purpose of testing it.[61]

At the same time Logan sent a description of the instrument to London to the Royal Astronomer, Edmund Halley. No acknowledgment was made, and in 1734 Logan sent a second description to Sir Hans Sloane and to Peter Collison for forwarding to the Royal Society. The arrival of this description coincided with the submission of the description of a similar instrument to the Society by its vice president, James Hadley. The Royal Society decided in favor of both inventors, and Godfrey was awarded the equivalent of 200 pounds in household furniture.

[60] "Frederick A. Heisely, Watch and Clockmaker and His Recorded Years, 1759–1839," *Timepieces Quarterly* (November 1948), vol. 1, no. 1, p. 33.

[61] Hindle, op. cit. (footnote 6), pp. 22, 68.

Figure 29.—Brass surveying compass made by Isaac Chandlee (1760–1813) of Nottingham, Maryland. Photo courtesy Ohio State Museum.

Godfrey is often confused with his son, also named Thomas Godfrey (1736–1763), who worked as a watchmaker in Philadelphia, and subsequently became active in literary arts.

Benjamin Condy (fl. 1756–1792, d. 1798) was an instrument maker with a shop on South Front Street in Philadelphia. As early as 1756 he worked for most of the merchant shippers of the port, supplying them with a considerable number of sand glasses that ranged from the quarter-minute to the two-hour varieties. Although he made his own mathematical instruments, it is likely that he imported the sand glasses. According to Customs House clearances of 1789, he had imported from London on the ship *Pigou* "three cases of merchandise" valued at £160/17/6 with a duty of $32.19, which may have included sand glasses.[62]

When Condy retired in 1792 he was succeeded in business by Thomas Biggs at the same address. Biggs had originally served an apprenticeship with Condy, and then fought for the American cause in the Revolution for five years. Following the termination of his military service he had engaged in instrument making in New York for eight years before returning to Philadelphia, his native city. Biggs prospered and his advertisements continued until early in 1795.

Thomas Pryor made instruments in a shop on Chestnut Street in 1778, but he evidently retired from business in the 1790's because the city directory of 1795 listed him merely as "gentleman." He is reported to have been one of those who, from the State House Yard, witnessed the transit of Venus.[63]

---

[62] GILLINGHAM, op. cit. (footnote 19), pp. 293–294.

[63] Ibid., p. 303; *Royal Pennsylvania Gazette*, April 19, 1778.

Among the early makers of mathematical instruments in Philadelphia was William Dean (?–1797), who is believed to have been working in that city as early as 1778. His name first appears in local directories in June 1792, where his shop address was listed as No. 43 South Front Street. Later he advertised that he made and sold "Surveying instruments—Telescopes, Sextants, Quadrants—and every article requisite for navigation, surveying, levelling, &c . . . ."

According to details which were noted in his last will, which was dated June 1, 1797, and filed and proved in the following month, Dean's death appears to have been preceded by a long illness. He designated his two sisters as his executrices, and the fact that his will specified the appointment of a Mr. Thomas Yardley, Jr., as guardian of his three children indicates that he may have been a widower at the time of his death.

A surveying compass by this maker was recently brought to light in the Clark County Historical Society, Springfield, Ohio, by Dr. Donald A. Hutzlar of the Ohio State Museum. The instrument is a plain compass in brass without levels, 13½ inches in length, and with a 5-inch needle. The dial is marked "DEAN PHILAD[a]." The wooden cover for the instrument is marked with the names of previous owners and dates, as follows:

Jno. C. Symes, Aug. 10, 1778
I. Ludlow, 1791
Henry Donnel, July 24, 1794
Jonathan Donnel, 1796
John Dyherty
Thomas J. Kizer, 1838
David J. Kizer, '78.

A description of this instrument in "*The History of Clark County, Ohio*" by A. P. Steele, published in 1881 by the W. H. Beers Co. of Chicago, adds considerably to its interest as a historical record of American scientific instruments and their use: "Col. Thomas Kizer, the veteran surveyor, has in his possession a compass made by Dean of Philadelphia; this instrument was owned and used by his father, David Kizer, who obtained it from John Dougherty about 1813; Dougherty got it from Jonathan Donnel. This relic is marked I. Ludlow, 1791; Henry Donnel, 1794; J. Donnel, 1796, John Dougherty, 1799; these marks are rudely scratched upon the cover of the instrument, and bear every evidence of being genuine; there is no doubt but that this old compass was used in making the first surveys in this county, or that it is the identical instrument used by John Dougherty, in laying off Demint's first plat of Springfield, and by Jonathan Donnel on the survey of 'New Boston.' " It is to be noted that some discrepancies exist in the listing of names and dates of the previous owners between Steele's *History* and those which actually appear on the cover of the instrument. Steele apparently made the changes he deemed necessary in his account of the instrument.

Between 1791 and 1795 the same address was also occupied by a cooper named Michael Davenport, and from 1797 to 1801 by "the Widow Davenport," presumably widow of Michael. From 1802 to 1804 the same address is listed for William Davenport, "Mathematical Instrument Maker," apprentice to William Dean, and believed to be the son of Michael. During the next ten years Davenport's address was 45 South Front Street, and then, to 1820, was 25 South Front Street.[64] Several brass surveying compasses bearing his name have survived.

Another maker of mathematical instruments about whom nothing further is known is Charles Taws, who was listed in this manner in the Philadelphia directory of 1795.

The making of instruments in glass appears to have been a specialized business in the Colonies, because those who worked in this field do not appear to have produced instruments in other materials. One of these makers of glass instruments—specifically barometers, thermometers and "Glass Bubbles to prove spirits, of different kinds"—was Alloysius Ketterer. He maintained a

[64] GILLINGHAM, op. cit. (footnote 19), p. 302.

Figure 30.—Brass surveying compass marked "F. Heisely Fred:*town." In collection of Ohio Historical Society, Ohio State Museum.

shop in the house of a Charles Kugler at "the sign of the Seven Stars," corner of Race and Fourth Streets in Philadelphia, in 1789. He moved to another address in Race Street in 1790 and was eventually succeeded in business by Martin Fisher, who increased the number of types of glass instruments made and sold at the shop.[65]

Henry Voight (1738–1814) was a man with a varied career. Of German ancestry, he was trained as a clock- and watchmaker, and he was a skilled mechanic. He operated a wire mill in Reading, Pennsylvania, in 1780 and moved shortly thereafter to Philadelphia, where he established a clockmaker's shop on Second Street. He became a close friend of the inventor John Fitch in about 1786, and in the following year he became a shareholder in Fitch's company for producing steamboats. In 1792 he entered into a short-lived partnership with Fitch to manufacture steam engines. In 1793 he invented a process for making steel from bar iron. In the same year President Washington appointed Voight to the position of chief coiner of the Philadelphia Mint, and he continued in that position until his death in 1814. He was closely associated with David Rittenhouse, Andrew Ellicott, Edward Duffield, and others.

Although there is no record of Voight's career as an instrument maker, there is nevertheless some evidence that he worked in that field. In the collection of the U.S. National Museum there is a brass equal-altitude telescope (fig. 31) made about 1790, that is signed "Henry Voigt." His name was spelled "Voigt" and "Voight" interchangeably.

Henry's son Thomas Voight worked as a clockmaker on North Seventh Street in Philadelphia around 1811. He was the maker of a tall case clock, ordered by Thomas Jefferson, that Jefferson's daughter presented in 1826 to her father's physician, Dr. Dunglison, for settlement of medical services.[66]

There were several instrument makers in provincial Pennsylvania, but the majority of such craftsmen worked in Philadelphia. Dr. Christopher Witt (1675–1765), an emigrant from England, worked in Germantown from about 1710 to 1765. He was well known locally as a medical doctor, scientist, "hexmeister," clockmaker, and teacher. It is traditionally claimed that he produced mathematical instruments in addition to timepieces. He described the great comet of 1743 and built his own 8-foot telescope. One of

[65] Ibid., pp. 305–306.

[66] Eckhardt, op. cit. (footnote 9), p. 195; George Evans, *Illustrated History of the United States Mint* (Philadelphia: Evans, 1890), p. 114.

his apprentices may have been Christopher Sower (1693–1740), of Germantown and Philadelphia, who achieved renown as a doctor, farmer, author, printer, papermaker, and clockmaker. He also produced mathematical instruments.[67]

George Wall, Jr., of Bucks County, was the author of a pamphlet on the subject of "a newly invented Surveying Instrument, called the Trigonometer." The instrument was described and illustrated in the pamphlet, which was published in Philadelphia in 1788. Washington's own copy, bearing the inscription "To the President of the United States from the Author" is in the collection of the Boston Athenaeum.

George Ford of Lancaster maintained a shop on West King Street, probably from the end of the 18th century until 1840. There he made tall case and other clocks, surveying compasses, and other instruments for the retail trade. However, he "did not push the business of Watchmaking and Clockmaking so hard, for the manufacture of nautical instruments and surveyors instruments was a more important part of his business." [68] Upon his death in 1842 he was succeeded by his son George Ford II.

Thomas Mendenhall repaired clocks and mathematical instruments in a shop on King and Queen Streets in the borough of Lancaster in 1775.[69]

John Wood of Philadelphia was a wholesale supplier of parts for clockmakers and watchmakers. According to a notice in the May 7, 1790, issue of *Pennsylvania Packet*, he had "pocket compasses, steel magnets, Surveying compass needles, surveyors chains, etc." Since no mention was made of making or mending instruments, it is probable that Wood was merely importer and wholesaler.

Another instrument maker of Philadelphia about whom little is known is Bryan Gilmur, who worked at the close of the 18th century making instruments and, possibly, clocks.[70]

James Jacks (also listed as James Jack) first worked as clockmaker and watchmaker in Charleston, South Carolina, in the 1780's; he later moved to Philadelphia where he maintained a shop

---

[67] Carolyn Wood Stretch, "Early Colonial Clockmakers in Philadelphia," *Pennsylvania Magazine* (July 1932), vol. 56, pp. 225, 235; Eckhardt, op. cit. (footnote 9), pp. 18, 24, 198.

[68] D. F. Magee, "Grandfather's Clocks: Their Making and Their Makers in Lancaster County," Papers read before the Lancaster (Pa.) Historical Society, 1917, pp. 63–77.

[69] Prime, op. cit. (footnote 17), p. 260.

[70] Palmer, op. cit. (footnote 34), p. 200.

Figure 31.—Equal altitude telescope, 17 in. long, made and signed by Henry Voight (1738–1814) of Philadelphia. USNM 311772.

on Market Street where he sold a variety of instruments. In the June 5, 1797, issue of *The Federal Gazette* he announced that, in addition to jewelry, clocks and watches, he "also had for sale mathematical instruments in cases very compleat; Surveyors Compasses and Theodolites; ship's Quadrants; Fishing Rods and Reels; Billiard Balls and sheet ivory; silver and plated coach, chaise and chair Whips."

Instruments of Wood

The Use of Wood

AN INTERESTING FACT concerning the instruments produced by 18th-century craftsmen is the relatively high incidence of instruments constructed of wood instead of brass or other metals. A significant reference to this use of wood is found in Alexander Hamilton's "Report on the Subject of Manufactures," published in 1821,[71] which refers to such items of wood as "Ships, cabinet-wares and turnery, wool and cotton cards, and other machinery for manufactures and husbandry, mathematical instruments," . . . and "coopers' wares of every kind."

Most common of these mathematical instruments is the surveying compass, possibly the instrument most needed and produced in America. Recorded in public and private collections are 31 known examples of such compasses made of wood, a rather large number. Furthermore, a substantial number of these were being produced simultaneously by skilled craftsmen who at the same time were making similar instruments in brass.

Finally, from a study of the surviving examples of wooden surveying compasses comes the interesting and perhaps significant fact that all the known makers were from New England. The towns and cities in which they worked were Boston and Plymouth in Massachusetts, Windsor and New Milford in Connecticut, and Walpole and Portsmouth in New Hampshire. A careful study of the advertisements and works of the instrument makers in the other large cities of the Colonies, such as New York, Baltimore, and Philadelphia, reveals no examples of wooden scientific instruments. Excluded, of course, are those instruments normally made of wood, such as the octant and the mariners quadrant.

Two possible exceptions are instrument makers of New York City. The first is James Ham, a maker of mathematical instruments "at the house wherein the Widow Ratsey lately lived

---

[71] ALEXANDER HAMILTON, *Official Reports on Publick Credit, A National Bank, Manufactures and a Mint* (Philadelphia: Wm. McKean, 1821), pp. 208–209

near the Old Dutch Church on Smith Street" who advertised in the May 27, 1754, issue of *The New York Mercury* that he made and sold

> mathematical instruments in wood, brass, or ivory, theodolites, circumferentors, sectors, parallel rules, protractors, plain scales, and dividers, the late instrument called an Octant, Davis' quadrants, gauging rods, sliding and gunter's scales, amplitude wood box and hanging and pocket compasses, surveying chains, japanned telescopes, dice and dice boxes, mariners compasses and kalenders, etc.[72]

Ham subsequently moved his business to Philadelphia where he first advertised in 1764, stating that he worked at the sign of "Hadley's Quadrant" at Front and Water Streets in Philadelphia and sold all forms of instruments in silver, brass, and ivory as well as "large brass pocket dials, fitted to the latitude of Philadelphia." In 1780 his son James Ham, Jr., advertised from the same address as a maker of mathematical instruments, specializing in "Hadley and Davis Quadrants." [73]

The second exception is William Hinton, who advertised in *The New York Gazette and the Weekly Mercury* of May 4, 1772, as follows:

> WILLIAM HINTON, Mathematical Instrument Maker, at Hadley's Quadrant, facing the East Side of the New Coffee House, Makes and sells all sorts of Mathematical Instruments, in Silver, Brass, Ivory or Wood, viz. Hadley's Quadrants, Davis's do. Crostaf's Nocturnals, Gunters Scales, Plotting do. Cases of Instruments, Surveyors Chains, Dividers with and without Points, Protractors, paralelled Rulers, Rods for Guaging, Amplitude, hanging and common Wood Compasses, Pocket do. three Foot Telescopes, Pocket do. Backgammon Tables, Dice and Dice Boxes, Billiard Balls and Tacks, Violin Bows and Bridges; with a Variety of other Articles too tedious to mention: And as he is a young Beginner, he flatters himself, he shall meet with Encouragement; and all those who please to favour him with their Custom, may depend upon having their Work done in the neatest and best Manner, and at reasonable Rates.

It is mentioned that both Ham and Hinton worked in wood in addition to other materials, but it appears very likely that the use of wood referred specifically to those instruments normally made of wood, such as quadrants and octants, and not to other instruments.

Any attempt to relate the making of wooden scientific instruments with the production of wooden clocks in New England has no conclusive result, yet there appears to be some relationship

[72] Rita Gottesman, *The Arts and Crafts in New York, 1726–1776* (New York: New York Historical Society, 1938), p. 307.

[73] Gillingham, op. cit. (footnote 35), p. 295.

between the two. Wooden clocks were made as early as the 17th century in Germany and Holland, and they were known in England in the early 18th century. In the Colonies the wooden clock was first produced in Connecticut, and the earliest type was associated with Hartford County. This form was quite common in East Hartford in 1761, and its first production may have had some association with Ebenezer Parmele (1690–1777), since an association between Parmele and all of the earliest makers of wooden clocks can be traced.[74] Little is known about Parmele. His father was a cabinetmaker in Guilford, Connecticut, and Ebenezer practiced the same craft, in addition to being a boat builder. He was a man of means, held various town offices, and served as town treasurer. For a while he operated a cargo sloop on Long Island Sound. In 1726 he built the first tower clock in Connecticut for the Guilford meeting house. He was a versatile worker in wood, and it is believed that he served an apprenticeship in New York City with a Dutch clockmaker from 1705 to 1710, where he may have learned to make wooden clocks.

This early type of wooden clock is associated with Benjamin Cheney (1725–1815), a clockmaker of East Hartford. The early or "Cheney" type of wooden clock was produced in Connecticut as late as 1812. A later form of the wooden movement began to appear about 1790, and was probably introduced by Gideon Roberts (1749–1813) of Bristol. Roberts had lived in the Wyoming Valley of Pennsylvania before 1790, and it is conjectured that he became familiar with the wooden clocks produced by the German settlers of that region.[75]

It is not surprising that the wooden clock had its colonial origins in Connecticut, so completely was it adaptable to the pioneer conditions in that colony. The materials were the abundant native woods—cherry, apple, oak, and laurel. The parts were made with simple carpenter tools and a wooden foot lathe, using the methods of the cabinetmaker. Although it has been suggested that some relationship may have existed between the makers of wooden instruments in England, and the makers of wooden clocks and scientific instruments in the New England Colonies,[76] a careful study has failed to reveal any connection,

---

[74] Hoopes, op. cit. (footnote 50), p. 3; and Hoopes, op. cit. (footnote 24), pp. 101–103.

[75] Hoopes, op. cit. (footnote 19), pp. 106–107.

[76] E. G. R. Taylor, *The Mathematical Practitioners of Tudor and Stuart England* (Cambridge: Cambridge University Press, 1954), pp. 185–292.

and there appears to be little if any parallel between the two groups. Basically, the use of wood for making some mathematical instruments in New England resulted from the native familiarity with this material, which was also employed to a considerable degree for the construction of domestic and agricultural implements, and from the fact that many of the early clockmakers had been trained as or by cabinet makers, carpenters, and even dish turners. Random examples of a few of the more prominent clockmakers are Joseph Hopkins, a wood turner; Chauncey Jerome, who had been apprenticed to a wood turner; and Silas Hoadley, who had worked with a cabinet maker.

Perhaps a basis for the prevalence of wood in these trades is to be found in the lines from a familiar poem:

> The Yankee boy, before he's sent to school,
> Knows well the mystery of that magic tool,
> The Pocket knife.[77]

But, from the technical point of view, it should be noted that those craftsmen who produced clocks and instruments and did not have their own brass foundries probably found that a good piece of straight-grained hardwood was as stable for holding its dimensions with the grain as a piece of brass. Shrinkage was at right angles to the grain; hence, for fixed linear stability wood was as good as brass. For rigidity per unit weight, wood was better than brass; and for availability and ease of working, wood was superior to brass.

It has often been ventured that wooden clocks were first produced in Connecticut, because of the scarcity of brass for this purpose during the years between the beginning of the Revolution to the end of the War of 1812. The claim is made that brass was not being produced in the Colonies and that it was imported exclusively from England during this period. Certainly, the wholesale price index of metal and metal products shows a steady increase during this period, and a considerable jump during the period of the War of 1812, making brass an extremely expensive material. This may explain why the makers of clocks and instruments who made and sold brass clocks and instruments were producing the same products at the same time in wood which, as we have seen, was both plentiful and a satisfactory substitute.

It can be surmised, therefore, that surveying instruments, as well as instruments for other purposes, were produced in both brass

---

[77] John Pierpont, "Whittling, A Yankee Portrait."

and wood simultaneously by many of the New England makers in order to provide suitable instruments in a flexible price range to meet the demands of the trade. Whereas today modern manufacturing methods make it possible to produce instruments in a wide variety, both in quality and price, to suit the needs and capabilities of every prospective purchaser, the production facilities of the 18th century were much more limited. The constant factor of skilled hand labor was costly. Metal was expensive. As evidenced in the records of Daniel Burnap, for instance, it was possible to produce surveying compasses in brass in two grades, presumably one more elaborate than the other. Yet Burnap's prices ranged between six pounds and four pounds for the metal instruments, making them still well out of reach of many of the would-be surveyors. Accordingly, Burnap—and presumably numerous other instrument makers of the period—produced from wood an economy model that sold for not more than two pounds, thus placing the item within the reach of the nonprofessional surveyor.

This theory is supported amply by the discovery that several of the instrument makers who worked in brass also made instruments of wood during the same periods. In addition to the evidence in the records of Daniel Burnap, there are the surviving surveying instruments in brass and wood made by Samuel Thaxter, Thomas Greenough, and John Dupee, leaving little if any doubt that the reason for producing surveying compasses and similar items of wood during the 18th century was to satisfy the need for reasonably accurate yet inexpensive instruments.

Surviving Instruments

The fact that the surviving examples of the wooden instruments were produced only in New England seems to indicate merely that the New England instrument makers were more familiar with the use of wood as a material, and had greater facility in working with it.

Undoubtedly other instruments produced by the 18th-century American makers have survived in addition to those already found. Quite likely examples of these wooden instruments still remain hidden in unexplored attics and other repositories. Yet, if the few thus far discovered is any criterion, the number ultimately recoverable will probably be but a fraction of the great number produced by the 18th-century makers during the half century or more

Figure 32.—Wooden graphometer used by Rev. Eleazar Wheelock (1711–1779) about 1769 for surveying the area of Dartmouth College in Hanover, New Hampshire. The hardwood block is covered with a brass plate with brass sighting bars mounted on a swivel and a spirit level under a brass strip on edge of instrument. The instrument is 8⅝ in. long, 4⅝ in. wide, and 7/16 in. thick. In collection of Dartmouth College Museum.

in which they worked. Even allowing for those probably destroyed in the natural course of events, one cannot help but wonder what has happened to the remainder.

A list of the surviving wooden instruments is given in the Appendix (p. 153). Many of these wooden instruments bear signatures or other marks that permit identification of their makers, but a number of specimens have been found that are not signed. In most instances they show evidence of professional workmanship, and they may have been the work of known craftsmen. One or two examples are obviously homemade by unskilled amateur practitioners.

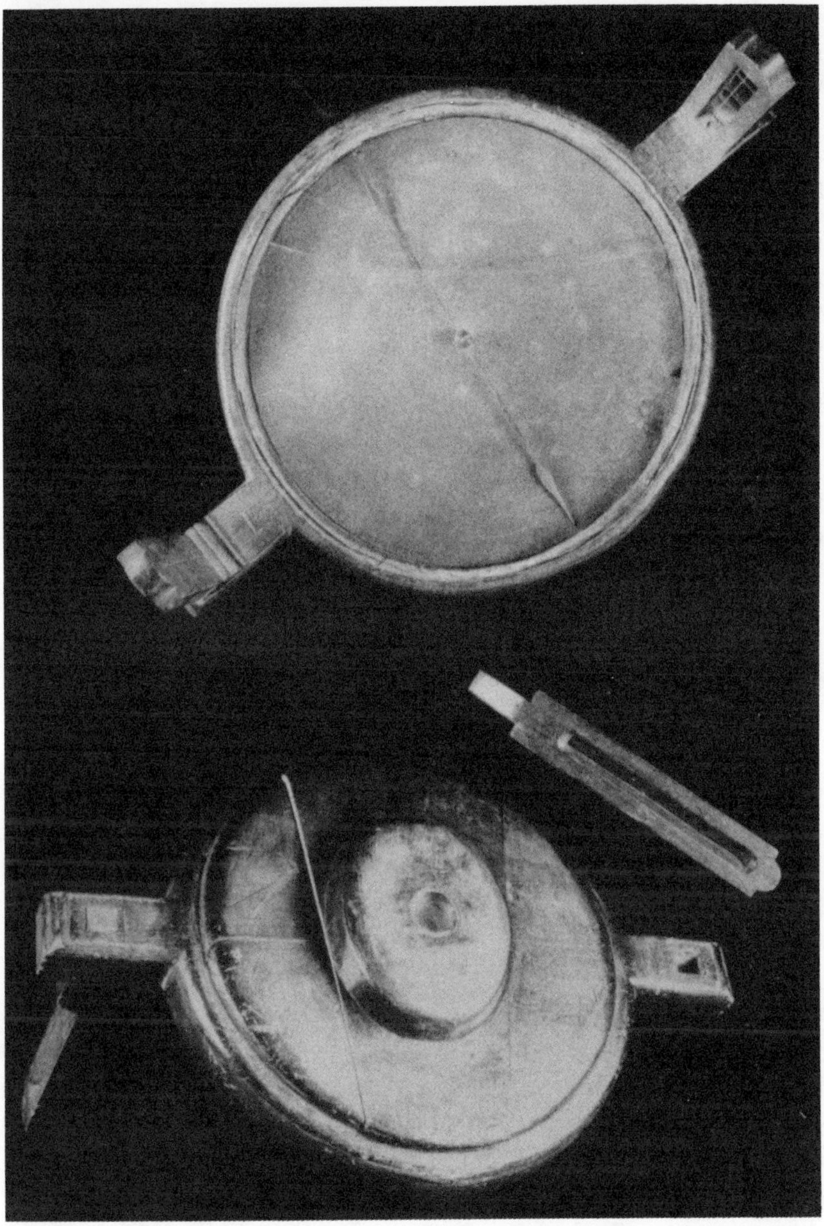

Figure 33.—Wooden surveying instrument, maker not known. Compass dial is of metal, painted green, with degrees marked to 90° with metal punches and the letter "N" to designate the north point. The instrument is 12 in. long; diameter is 8 in. In collection of Dartmouth College Museum.

Several unsigned wooden instruments of professional quality are in the collection of the Dartmouth College Museum. Of particular interest is a semicircumferentor (fig. 32) that belonged to the Reverend Eleazar Wheelock (1711–1779) who founded Moor's Indian Charity School at Lebanon, Connecticut, which subsequently developed into Dartmouth College. It is claimed that it was with this instrument that the area of the college was surveyed when it moved to Hanover, N.H. The instrument is actually a graphometer consisting of a block of hard wood faced with a brass plate with a trough compass; it is tentatively dated about 1769. The identity of the maker is unknown, but it may have been the product of Hagger, who made a similar instrument, illustrated here, or it may have been produced by any one of the other makers noted. The type of instrument is an old one. It is described in John Love's *Geodaesia, Or the Art of Surveying and Measuring Land*, published in London in 1688. Abel Flint[78] also commented on this semicircle as being sometime used, as well as the plane table and perambulator—

> . . . but of these instruments very little [use] is made in New England; and they are not often to be met with. For general practice none will be found more useful than a common chain and a compass upon Rittenhouse's construction.

Another of the unusual wooden surveying instruments in the collection of the Dartmouth College Museum is a wooden surveying compass (fig. 33) in which the sighting bars appear relatively close to the dial. A metal plate, painted green, is stamped with the degrees marked to 90°. A single N for the north point is stamped into it, presumably with steel punches. The instrument is relatively primitive, and is sufficiently different from the other examples noted to merit mention. There is no maker's name, nor any clue to the date or place or period of origin.

An unsigned semicircumferentor made of wood is owned by Mr. Roleigh Lee Stubbs of Charleston, West Virginia. The instrument measures 3¾ in. by 7½ in. by 1 in., and there are sighting bars 3 in. high on a swinging brass bar pinned at the center of the base. It has a trough compass, and the gradations around the edge of the semicircle are marked with tiny brass pins. The date "1784" is stamped into the wood with the same type of figures as appear in the degree markings, probably with small steel punches.

A surveying compass of the conventional type, also made of

---

[78] Abel Flint, *System of Geometry and Trigonometry together with a Treatise of Surveying* (Hartford: Olive D. Cooke, 1804), p. 86.

Figure 34.—18th-century semicircumferentor. Inscribed brass plate is mounted on a mahogany block; brass sighting bars are mounted on a swivelling bar. The trough compass is on a silvered dial. In collection of the writer.

wood, is in The Farmer's Museum at Cooperstown, New York. The wood is ash or oak, 12¾ in. long and 6½ in. in diameter, with the sighting bars 5 in. high. The compass card consists of cut-out printed letters pasted upon a printed compass rose, and the fleur-de-lis at North is inked-in by hand. This may be a homemade replacement of the original card. The instrument is believed to date between 1760–1775.

Of equal interest is a large semicircumferentor made by an unknown American instrument maker in the second half of the 18th century. The instrument (fig. 34) consists of a plate of hammered brass attached to a quarter circle block of mahogany, with a glass covered trough compass within a silvered opening, and the gradations stamped into the brass. The brass sighting bars are attached to a swivelling bar that can be fixed in place with a set screw underneath the block. The instrument, which is in the collection of the writer, is not signed with a maker's name. Its workmanship is excellent, and professional.

On the basis of a comparison of these instruments with those produced by known professional makers, it becomes apparent that all of them were made professionally. The possibility that some of these wooden surveying compasses may have been produced by the farmer or local surveyor for his own use is extremely unlikely. Homemade instruments such as those described below were unquestionably the exception instead of the rule.

Figure 35.—Homemade wooden surveying compass carved from block of maple entirely with a jackknife; painted in red In collection of Preston R. Bassett, Ridgefield, Connecticut.

An exception to this generalization, and an extremely fine example of the whittler's art, is a surveying compass (fig. 35) in the collection of Mr. Preston R. Bassett of Ridgefield, Connecticut. This is a comparatively small instrument made of maple; the body was painted red. It is carved entirely by means of a jackknife, and the sighting bars are also whittled to shape and mortised permanently into the frame. A lid covering the dial is carved from soft pine. The compass dial is handdrawn in black ink, and the North point is painted in the form of a decorative fleur-de-lis in red and green. A homemade ring of pewter surrounds the compass rose at needle level. This is graduated in degrees, with every 10° marked, stamped with steel punches. The ring is set into the base by means of wooden pegs. The steel needle is nicely cut, and it is probably the only part purchased by the maker.

This is unquestionably a homemade instrument produced by a skillful whittler early in the 18th century.

Compass Cards

A fact that becomes apparent in a comparison of the surviving examples of wooden surveying compasses made in New England is the similarity of the compass cards used by makers in the seaport cities (see fig. 36). The compass card in each of these instances is the type designed for a mariner's compass, bearing a star of 32 rays to mark the 32 points of the heavens. The North point is designated with an elaborate fleur-de-lis, and the East is emphasized with scrollwork. These are features which were not designed primarily for land surveying. Presumably, these makers had a quantity of engraved or printed compass cards that they used in both marine and land surveying compasses. This is true in the case of the compasses made by James and Joseph Halsy, Greenough, Clough, Warren, Thaxter, Dupee, Breed, and Bowles. On the other hand, the dial of Huntington's compass was painted directly on the wood, and the semicircumferentors do not utilize the marine compass card. Obviously these makers resorted to this practice for reasons of economy—to reduce costs of engraving and printing, and using the same card for both types of instruments that they produced.

Trade Signs

An interesting sidelight in the study of the makers of scientific instruments is the advertising they used, particularly the design

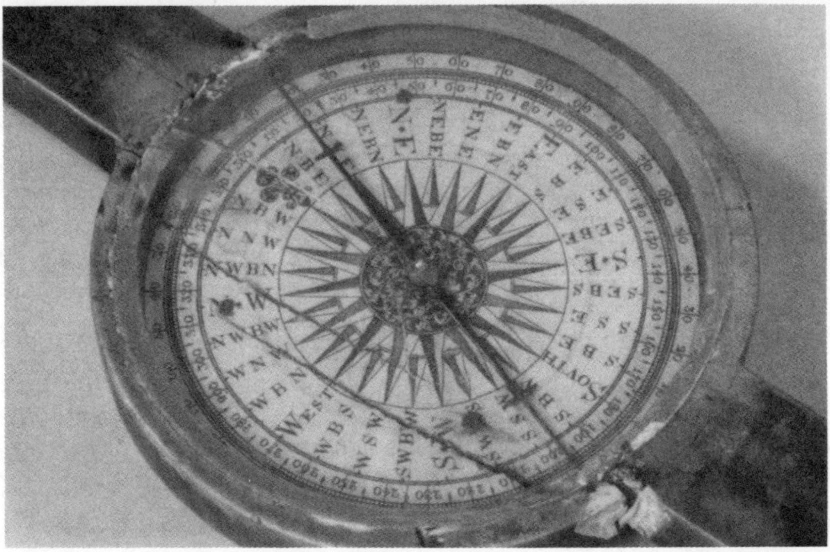

Figure 36.—Unsigned wooden surveying compass, with an interesting example of a mariner's compass card.

of their signboards. The most popular symbol appears to have been the quadrant, as the phrase "At the Sign of the Quadrant" is found repeatedly in advertising in several of the seaport cities of the 18th century.

In Providence, William Hamlin used the designation in the first part of the 19th century, while Philadelphian John Gould featured the sign at the end of the 18th century. During an even earlier period, William Hinton designated his address to be "At Hadley's Quadrant" in New York City. Both Gould and Hinton were English, which may have had some bearing on their selection of the quadrant as a symbol of their merchandise.

Other signboards were as colorful, such as Jonathan Dakin's "Sign of the Hand and Beam," James Youle's "Sign of the Cross-Knives and Gun," and Charles Kugler's house in Philadelphia with its "Sign of the Seven Stars" (that is, Great Bear), which housed the shops of several instrument makers.

The two most interesting and significant of the instrument makers' trade signs were those advertising the shop of Samuel Thaxter. The first of these was the carved wooden figure of "The Little Admiral," which was a favorite landmark at No. 1 Long Wharf in Boston for almost a century and a half. It was

Figure 37.—"The Little Admiral," trade sign used for almost a century and a half in Boston, first by William Williams and later by Samuel Thaxter. Reputed to have been carved by John Skillin of Boston. In collection of the Bostonian Society.

the handiwork of John Skillin, the 18th-century woodcarver of Boston, upon whose death on January 24, 1800, the *Chronicle* commented that "he was for many years the most eminent cf his profession." John Skillin and his brother Simeon worked in Boston from about 1777 and produced most of the figureheads that issued from that port during that period, as well as a number of other notable ornamental wooden figures.

According to Mrs. H. Ropes Cabot of the Bostonian Society, the figure of "The Little Admiral" (fig. 37) had been carved for William Williams, who brought it with him to Boston from Marblehead in 1770 when he established his shop. The figure was installed in front of the Crown Coffee House, and Williams's shop was thereafter designated by this symbol. The trade sign survived through the years of the Revolutionary War. When the original building of the Coffee House was burned, the carving was saved and installed on the new building erected in its place. In an account of Boston landmarks, Porter [79] related the figure to the Admiral Vernon Tavern at the eastern corner of Merchants Row. He was proved to have been in error, however, since the trade sign of that public-house was a portrait bust of Admiral Vernon and the place was known as the Vernon Head Tavern for half a century, even after the end of the Revolution.

When Samuel Thaxter purchased the business from Williams's estate he acquired the figure as well, and he moved it to each new location for his shop. The figure of "The Little Admiral" continued to designate the firm even after Thaxter's death, until the firm finally went out of existence at the beginning of the 20th century. When the old store was torn down in 1901, the figure was preserved, presumably by the last owner's family. In 1916 it was acquired for the Bostonian Society by several of its members, and the figure has been preserved in the Society's Council Chamber since that time.

The other interesting trade sign utilized by Samuel Thaxter is a carved figure of Father Time that is credited to John Skillin (see fig. 38). The figure is believed to have been commissioned by Thaxter during the last decade of the 18th century and installed by him in the interior of his shop. It is an important example of the American woodcarver's art, and is equivalent to the best work of the Skillin brothers.

---

[79] "Report of the Committee on the Rooms," *Proceedings of the Bostonian Society* (1917), no. 1, p. 16.

Figure 38.—"Father Time" trade sign used by Samuel Thaxter in his shop in 18th and 19th centuries. Made of wood, it was carved by John Skillin of Boston. In collection of the Bostonian Society.

The Makers

Surprisingly, the names of the craftsmen who produced wooden instruments are not noted among the instrument makers. With only one or two exceptions, their names are hitherto unknown in the history of American science, and for that reason it has been considered advisable to present all available information that could be accumulated about them.

Joseph Halsy

The earliest known maker of wooden scientific instruments of Boston was Joseph Halsy. He appears to have been one of the sons of the James Halsie I, who was mentioned in a land deed of 1674 as a mathematician.[80] The land records indicate that James I was the father of several children, including Rebecca, a spinster; John Halsey, a mariner who died before 1716; Sarah, who later became Mrs. Dorsan; another daughter, name unknown, who became the wife of a Joseph Gilbert and the mother of two daughters and a son who inherited her share of her father's estate; Nathaniel Halsie; and probably Joseph Halsy. James Halsie I appears to have owned property consisting of land, a wooden house, and wharves on the North End, on North Street between Sun Court and Fleet Street.[81]

The date of birth of Joseph Halsy of Boston has not been found, but mention is made of the fact that on January 29, 1697, he was married to Elizabeth Eldridge, the daughter of a mariner named Joseph Eldridge, and that five children resulted from the marriage, three sons and two daughters.[82] One son, Joseph, died in infancy and a daughter, Elizabeth, died at an early age.

On February 26, 1704/5 Halsy purchased from Rebecca Halsey, the spinster daughter of James, her share in the house and land of her late father on North Street between Sun Court and Fleet Street.

On April 19, 1714, Halsy and his wife deeded a house and land on North End, at the corner of Hanover and Salutation Streets, to a shipwright named Joseph Hood. Two years later, on March 2, 1716, he purchased from Jane, his sister-in-law, who was the widow

[80] Savage, op. cit. (footnote 2), vol. 2, p. 341.

[81] "James Halsy," in Thwing Catalogue, Massachusetts Historical Society.

[82] Savage, op. cit. (footnote 2), vol. 2, p. 341.

of the mariner John Halsy, her share of the house and land of James Halsie, being the same property on North Street. On March 27 of the same year he purchased the share in the same property belonging to Sarah Dorsan, his widowed sister. In August 1719 he was forced to mortgage some of the property to a merchant named John Frizell, but the mortgage was cancelled in 1741.

Halsy was married for a second time on January 10, 1731, to Mrs. Anna Lloyd, a widow.[83]

During the 1730's, Halsy continued to buy out the heirs of James Halsie. On March 6, 1730, he acquired the share of Mary Gilbert, a granddaughter, and on the same date he purchased from the James Halsey heirs their inheritance "part to land, wharf, house, shop and buildings on North Street." Other heirs remained, for in June 9, 1732, he bought out the share of Marty Partridge, another granddaughter, and on June 27 the share of Joseph Gilbert, Jr., a grandson. In October 1740 he was forced to mortgage as security to James Bowdoin a house and land on the southwest side of North Street, but this was cancelled when on August 26, 1751, Joseph Halsey and his wife, Anna, deeded to James Noble the land, wooden house, and wharves near Fish Street on North Street between Sun Court and Fleet Street, which apparently was

[83] Ibid.

Figure 39.—Wooden surveying compass "Made and sold by Joseph Halsy, Boston, New England." The instrument, made of maple, is 11 in. long and has a diameter of 5¾ in. In the collection of New Hampshire Historical Society, Concord.

formerly the property of James Halsey that Joseph had acquired with so much trouble over a period of 40 years.[84]

The following advertisement relating to instruments sold by Halsy appeared in the issues of *The Boston Gazette* for the months of September and October 1738:

> Made and sold by Joseph Halsey jun. Hadley's New Invented Quadrant or Octant the best and exacted Instrument for taking the Latitude or Other Altitudes at Sea, as ever yet Invented.[85]

The last dated record relating to Joseph Halsy which has been found is a letter dated February 3, 1762, that he wrote to Robert Treat Paine concerning legal matters.

Only one complete instrument produced by Joseph Halsy appears to have survived—an especially fine wooden surveyors compass (fig. 39) in the collection of the New Hampshire Historical Society. It is made of maple. The compass card, probably the most interesting of any found in the wooden instruments, is hand-colored in black, blue, red, and gold. A fleur-de-lis marks the North point, and triangular pointers indicate the other compass directions. Inside the pointers are crudely painted female figures representing the seven arts: NW, Grammar; W, Logick; SW, Geometry; S, Arithmetick; SE, Astronomy; E, Rhetorick; and NE, Musick. Within a medallion at the center of the compass card is depicted a sailing vessel at sea; surrounding the medallion is a riband inscribed "Made and Sold by JOSEPH HALSY Boston—New England." [86]

Another, but much less elaborate, compass card used by Joseph Halsy, is an engraved example (fig. 40) found glued in Thomas Paine's own manuscript copy of Charles Morton's *Compendium Physicae*, which is preserved in the collection of the Massachusetts Historical Society.

John Halsy (fl. 1700–1750), also a mathematical instrument maker, had a shop on Green Street, in Boston, according to the Record Commissioner's "Report of the City of Boston." He was married on December 10, 1700, by the Reverend Cotton Mather. He probably was a brother of Joseph Halsy who worked in the same period.

John Halsy subsequently abandoned his instrument-making

---

[84] "Joseph Halsy," in Thwing Catalogue, Massachusetts Historical Society.

[85] *Boston Gazette*, September 18–25, October 2–9, and October 16–23, 1738.

[86] Description courtesy of Mr. Philip N. Guyol, director, New Hampshire Historical Society.

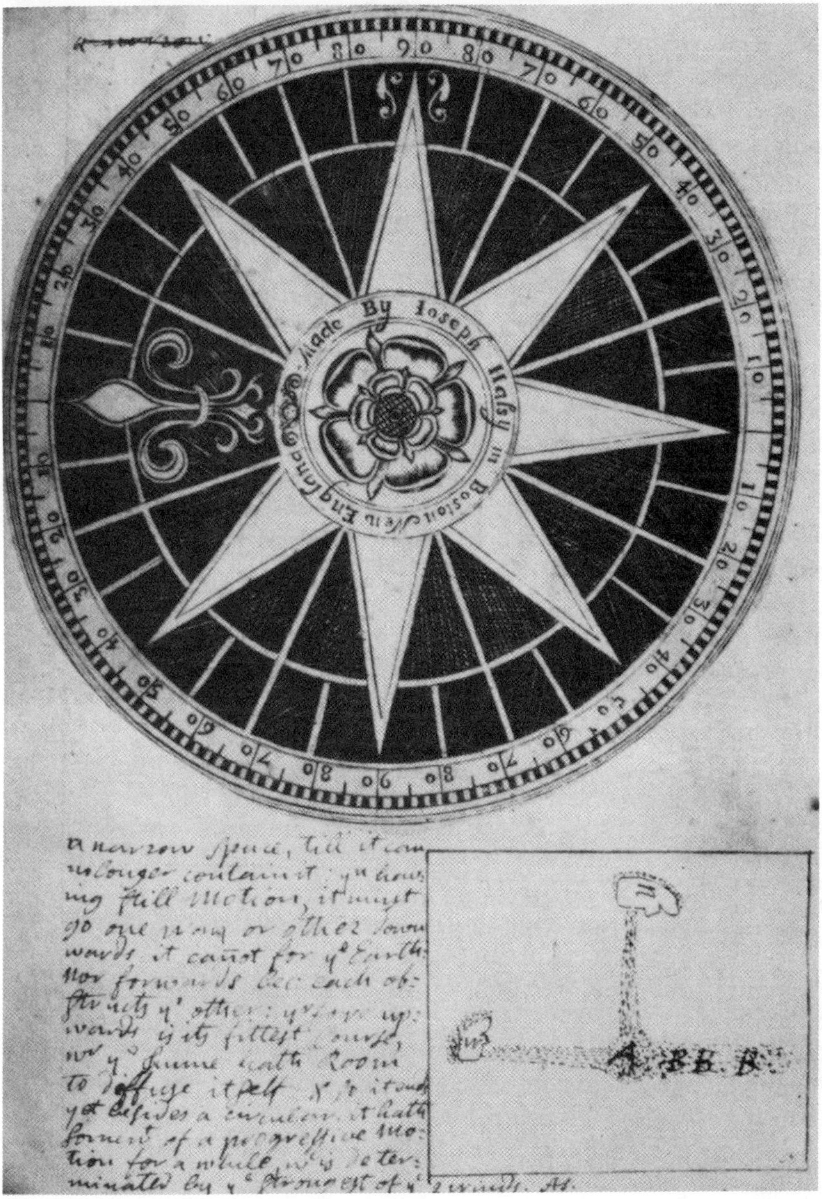

Figure 40.—Compass card of Joseph Halsy found glued into Thomas Paine's personal copy of Charles Morton's *Compendium Physicae*. In collection of Massachusetts Historical Society.

business to become a pirate. He went out to Madagascar, where it is reported that he died in his own bed. He was buried with the rites of the Church of England in his own watermelon patch.

James Halsy II

James Halsy II (1695–1767), a mathematical instrument maker, was born in Boston on April 10, 1695, the son of Nathaniel and Hannah (Gross) Halsie. The parents had been married by the Reverend Cotton Mather in June 1693.[87] In 1716 young James Halsy was a member of the Artillery Company, and by 1720 he had the rating of 4th sergeant. He held town offices and was one of the founders of the New Brick Church of Boston. On May 30, 1717, he married Anna Gutridge (Goodrich). Ten years later, on September 22, 1727, he bought a house and land on North Bennett and Tileston Streets from Hugh Hall, a merchant; at the same time he deeded to Hall some land and a house adjacent to the latter on the southwest side of Green Street. On January 5, 1837, he deeded to his aunt(?), a single woman named Huldah Gross, a house and land on Ann Street that he had inherited from Thomas Gross, his grandfather. Several more real estate negotiations were recorded in the course of the next few years. In October 1740 he purchased a house and land on the north side of North Bennet Street from John Endicott; in January 1741 land on the east side of North Bennett Street; and in November 1748 half of the house and land of Edward Pell, adjacent to Huldah Gross, on Cross Street; finally, in October 1753, he purchased land on Tileston and North Bennett Streets from John Grant.[88]

Halsy died on January 2, 1767, at the age of 72. In his will dated May 1, 1766, and probated January 2, 1767, by which his wife Anna was the executrix of his estate, he left her the income of his real and personal estate. He apparently was survived by three daughters and a son, also named James Halsy. He divided his real estate in Boston amongst his daughters, and to his son he left land in New Hampshire.[89]

The only known surviving instrument bearing James Halsy's name is a wooden surveying compass (fig. 41) in the collection

---

[87] SAVAGE, op. cit. (footnote 2), vol. 2, p. 341; "Joseph Halsy," in Thwing Catalogue, and "Cotton Mather" in Record of Marriages, Massachusetts Historical Society.

[88] Land deeds listed in Thwing Catalogue, Massachusetts Historical Society.

[89] Massachusetts Historical Society, Inventory L.450, S.P.R. 92.505.

Figure 41.—Wooden surveying compass made by James Halsy (1695–1767) of Boston. The instrument is 11 in. long. In collection of East India Marine Hall, Peabody Museum, Salem, Massachusetts.

of the Peabody Museum in Salem. The engraved compass card is quite similar to the one used by Thomas Greenough. In the central medallion is an elaborate royal crown, and in the circle around the medallion is inscribed "Made and Sold by JAMES HALSY near Ye Draw Bridge in Boston."[90]

Thomas Greenough

Contemporary with James Halsy II was Thomas Greenough (1710–1785), who was born in Boston in 1710, the son of John and Elizabeth (Gross) Greenough. His father was a shipwright in the North End of Boston, and one of Thomas's brothers, Newman Greenough, became a sailmaker. Thomas also had a sister named Jerusha, who later figured in his real estate negotiations.

The earliest known record relating to Greenough is of his marriage in 1734 to Martha Clarke, daughter of William and Sarah Clarke of Boston. Nine children resulted from this marriage over the course of the next 16 years; four of these were sons. On January 27 of the year of his marriage he purchased a house on the northwest side of North Street, between Mill Creek and Union Street, from John White and Nathaniel Roberts. On August 1, 1736, Greenough purchased the house and land of his father-in-law, William Clarke, on the south side of Portland Street. On

[90] Description courtesy of Mr. M. V. Brewington, Peabody Museum, Salem, Mass.

October 28 he mortgaged to his mother his house on Ann Street (which appears to have been the house he had purchased on North Street), and at the same time he deeded to his brother Newman all his right and title in his father's estate at the North End. Greenough was only 24 at the time of his marriage, and he apparently became involved in real estate, by choice or by necessity, to a considerable degree.

Greenough, in 1744, was a member of a militia company in Boston,[91] and three years later, in 1747, he was listed as third sergeant. He was a firm patriot, held a town office, and was a founder and deacon of the New Brick Church in Boston.

Greenough had a substantial interest in the holdings of his late father-in-law. For example, on August 11, 1744, he and his wife deeded to a merchant named James Pitts the seawall, or new wharf, "before the Town of Boston in the front and rear lying to the northward of King Street Pier, North Wharf and flats of James Bowdoin," all of which was part of the estate of his deceased father-in-law that apparently had been inherited by his wife. In the following year, on November 1, 1745, he purchased a house and land on Portland Street from his widowed mother-in-law and then on March 31, 1746, he and his wife deeded the same house and land to a merchant named Stephen Hall. Numerous other negotiations of the same nature are on record.

At some time between 1748 and 1750 Greenough's first wife, Martha, died, and in 1750 he married Sarah Stoddard. Three more children, all sons, resulted from this second marriage. His real estate negotiations continued full pace during the second marriage as during the first.[92]

Greenough's second wife preceded him in death, and Greenough died in 1785 at the age of 75. His will, probated on August 23, 1785, had been made on May 21, 1782;[93] it contained some interesting bequests:

> Executors: my two sons, David S. and William Greenough. Legatees: to the children of my son Thomas, deceased, Rachel, Ann, and Sally Greenough, £13.6.8 each. To their sister Betty £5. To the children of my son John deceased, 200 acres of land. I also give his eldest son John my silver can, fellow to the one I gave his father. To his sons Wm. and David, and to his

[91] Called the "r r Co.," which has not been further identified but is believed to have been one of the many militia companies that were formed in Boston during this period.

[92] "Thomas Greenough," in Thwing Catalogue, Massachusetts Historical Society.

[93] M.S. identified as Folio 495, Massachusetts Historical Society.

daughters, Sarah, Abigail, and Mehitible £5 each and the house they live in. My daughter, Sarah Edwards, £10 and a silver chafing dish. My daughter Martha Stone all my lands in the County of York, Cape Porpoise, and Wells, and my silver salver, and her son Thomas £5 and a silver porringer. My daughter Elizabeth Brooks £10 and a silver tea pot. My daughter Mary Savage £40 and to her son Thomas one silver porringer. To the children of my daughter Jerusha, deceased, Martha Clark Lepear and Sally Lepear each of them, £50, and a pair of salt shovels, and a pepper box, silver. All the rest of my estate to my two sons, David Stoddard Greenough, and Wm. Greenough. The late Shute Shrimpton Yeoman, Esq., left an estate to my late spouse Sarah, and to her children, in the Island of Antigua. In case my son David should have a legal possession of same, and Wm. no part, in that case I give my son David £100 and sundry pieces as per schedule amount to £63.11.3. All the rest of my estate to my son, William Greenough.

Of particular interest with relation to Greenough's business in instruments is the following advertisement that appeared on May 11, 1742, in *The Boston Gazette:*

To be sold by Capt. Cyprian Southack at his House near the Orange Tree and at Mr. Tho. Greenough's Mathematical Instrument Maker near the Draw Bridge, said Southack's Char[t]s of the Coast from Sandy Point of New York to Canso.

Invaluable for this study are Thomas Greenough's manuscript accounts that have survived in the collections of the Massachusetts Historical Society. The following itemized entries are selected from Greenough's business accounts over a period of two decades to provide data on the prices current in the second half of the 18th century for new instruments and for repairing others:

In Account with Thomas James Gruchy:

1754, April 27: 1 Compass for the Schooner *Sea Flour*........£0.8.0.
1758, Nov. 28: 1 Spyglass.................................£1.13.8.
1759, Jan. 25: Mending 3 Compasses for the Schooner *Susanna*..............................£0.6.0.

In Account with Nathaniel Bethune:

1760, August: A gauging rod...............................£0.6.0.
Mending a telescope.......................£0.3.0.

In Account with Captain McAndrew Mirick of Nantucket:

1772, March 21: For 2 compasses, 1 leaded..................£0.16.8.

In Account with Captain Roberson Crockett:

1773, April: For mending 2 Compasses....................£0.6.2.
For mending 1 Hanging Compass..............£0.3.2.

In Account with Captain Reworth of the Brig *Fortune:*

1774, March 30: For mending 2 compasses & Glasses...........£0.7.0.

In Account with Captain Thomas Godfrey:

1774, April 7: For 1 Telescope.............................. £0.8.0.

Other documents in the same collection indicate that Greenough's business interests were substantial and not limited merely to the construction of instruments. On July 31, 1769, Greenough's name appeared on the Boston Citizens' Non-Importation Agreement. Subsequently, on December 14, 1774, there is Greenough's signed receipt, with the amount left blank, stating that he had "REC'D. of Capt. Thomas Godfrey the Sum of—— in full for my Negro man Cuffes Shair in the Whaling Voige ——."

Greenough apparently was succeeded in business by his son William Greenough. Mr. Lawrence B. Romaine of Middleboro, Massachusetts, in 1939 described a wooden surveying compass with its own hand-whittled tripod made of oak which bore a compass card inscribed "Made by William Greenough, Boston,

Figure 42.—Brass surveying compass made by Thomas Greenough (1710–1785) of Boston. Compass face is mounted on main blade with two copper rivets. Screws for vanes and tripod mounting are hand cut, with wing nut ends. Sighting bars are 1/16 in. wide and 5¼ in. high; over-all length is 11⅞ in. and diameter is 5¼ in. Owned by Greenough family of Boston. Photo courtesy of Dr. Thomas Greenough.

Figure 43.—Wooden surveying compass, made and sold by Thomas Greenough. The instrument is made of gumwood and has a paper compass card; it is 13¼ in. long and has a diameter of 5¾ in. In collection of Franklin Institute, Philadelphia.

N.E."[94] The compass was protected by a pine cover that fitted closely between the sights. The present location of this instrument is not known, but it appears to be the only known example by William Greenough made of wood.[95]

In the Greenough family at the present time is a brass surveying compass (fig. 42) of fine quality and of the period before or during the American Revolution. The dial is finely engraved with a Tudor rose at its center, and around it is the inscription "Thomas Greenough Boston Fecit." The compass face is mounted to the main blade with two copper rivets. The holding screws for the vane and tripod mounting are rather crudely hand cut with wing-nut ends.[96]

---

[94] *The Chronicle* (Early American Industries Association), December 1939, vol. 2, no. 12, p. 96.

[95] Ibid.

[96] Description courtesy of Dr. Thomas Greenough, Cooperstown, N. Y.

Figure 44.—Wooden surveying compass made and sold by Thomas Greenough. Made of hickory, it is 11 in. long and has a diameter of 5½ in. Compass card is of paper. Allegedly, this compass was used by Joseph Frye for surveying his land grant in what is now Fryeburg, Maine, in 1762. Loaned to the U.S. National Museum by Laurits C. Eichner of Clifton, New Jersey. USNM 315001.

Five other surveying compasses made by Thomas Greenough are known, and all are made of wood: the one in the Franklin Institute is made of gum (fig. 43), one in Old Sturbridge is made of maple, one in the Bucks County Historical collection at the Mercer Museum is made of cherry, one owned by this writer is made of basswood, and one on loan to the U.S. National Museum from Mr. Laurits C. Eichner is made of hickory (fig. 44).

The compass at the Mercer Museum forms part of the surveyor's gear used to lay out the town of Weymouth, Massachusetts. The example in hickory on loan to the U.S. National Museum, as is usually the case with the compass cards of the Thomas Greenough instruments, has the central ring printed in gilt, and the inscription has turned black, making the inscription almost illegible. This specimen was owned by Joseph Frye, who was given a land grant in what is now Fryeburg, Maine, in 1762. He allegedly used this compass for surveying that land. In 1783 he assembled a manuscript book of tables (see fig. 45) for use in surveying for his son Joseph Frye, Jr. This manuscript also is part of the loan to the U.S. National Museum.[97]

[97] ROBERT P. MULTHAUF, "Early Instruments in the History of Surveying: Their Use and Invention," *Surveying and Mapping* (October-December 1958), pp. 401, 403.

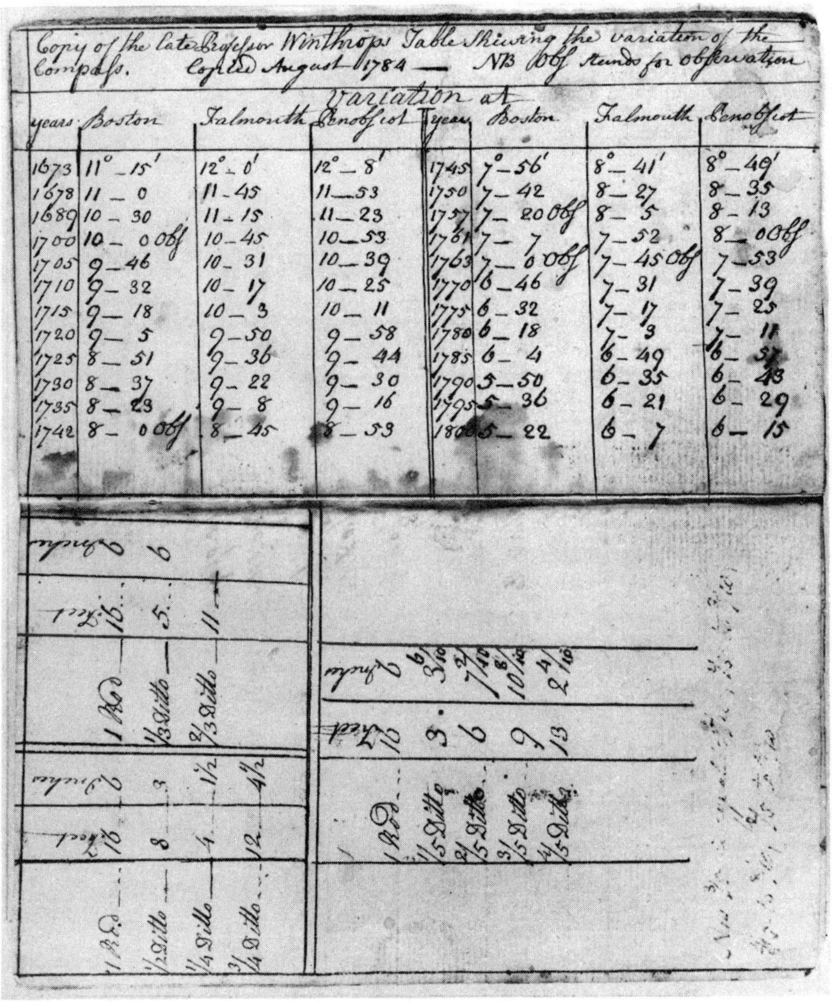

Copy of the late Professor Winthrop's Table Shewing the variation of the Compass. Copied August 1784 — NB Obs stands for observation

Variation at

years	Boston	Falmouth	Penobscot	years	Boston	Falmouth	Penobscot
1673	11°–15′	12°–0′	12°–8′	1745	7°–56′	8°–41′	8°–49′
1678	11–0	11–45	11–53	1750	7–42	8–27	8–35
1689	10–30	11–15	11–23	1757	7–20 Obs	8–5	8–13
1700	10–0 Obs	10–45	10–53	1761	7–7	7–52	8–0 Obs
1705	9–46	10–31	10–39	1763	7–0 Obs	7–45 Obs	7–53
1710	9–32	10–17	10–25	1770	6–46	7–31	7–39
1715	9–18	10–3	10–11	1775	6–32	7–17	7–25
1720	9–5	9–50	9–58	1780	6–18	7–3	7–11
1725	8–51	9–36	9–44	1785	6–4	6–49	6–57
1730	8–37	9–22	9–30	1790	5–50	6–35	6–43
1735	8–23	9–8	9–16	1795	5–36	6–21	6–29
1742	8–0 Obs	8–45	8–53	1800	5–22	6–7	6–15

Figure 45.—Pages from a booklet of "Tables Useful in Surveying Land, Made and presented by Joseph Frye to his son, Joseph Frye, Jr., November 18, A.D. 1783." Loaned to the U.S. National Museum by Laurits C. Eichner of Clifton, N.J. USNM 315062.

Figure 46.—Compass card from a wooden surveying compass "Made by Thomas Greenough, Boston, New England." In collection of the writer.

The compass card in each of these five instruments is identical, designed for use in the mariner's compass (see fig. 46). A gentleman in the dress of about 1740 stands on the shore using a Davis quadrant. Offshore in the harbor is a schooner of the 1750 period. Minor features of the scene are touched up in red, presumably printed, since they are consistent in all of the cards.

William Williams

Although not one of the earliest instrument makers in Boston, but certainly one of the more significant, was William Williams (1737/8–1792). He was the son of Capt. John Williams, a shopkeeper who died on March 22, 1748, at the age of 41, and who was buried in King's Chapel Burial Ground.[98]

William Williams was born in 1737 or 1738. He was ten years of age when his father died, and he had two brothers and two sisters. His father left a substantial estate of £6,575, of which £4,544/9/4 was for the inventory of the shop merchandise. One of the appraisers for his estate, Jotham Maverick, married the widowed Mrs. Williams less than a year later, on January 20, 1748/9.[99]

In 1770 William Williams established himself as a mathematical instrument maker and clockmaker at No. 1 Long Wharf, at the Crown Coffee House, as it was then known. The shop was located on the corner of State and Chatham Streets, on premises owned by Robert Shillcock.

---

[98] "Report of the Committee on the Rooms," *Proceedings of the Bostonian Society* (1917), no. 1, p. 14.

[99] Ibid., p. 15.

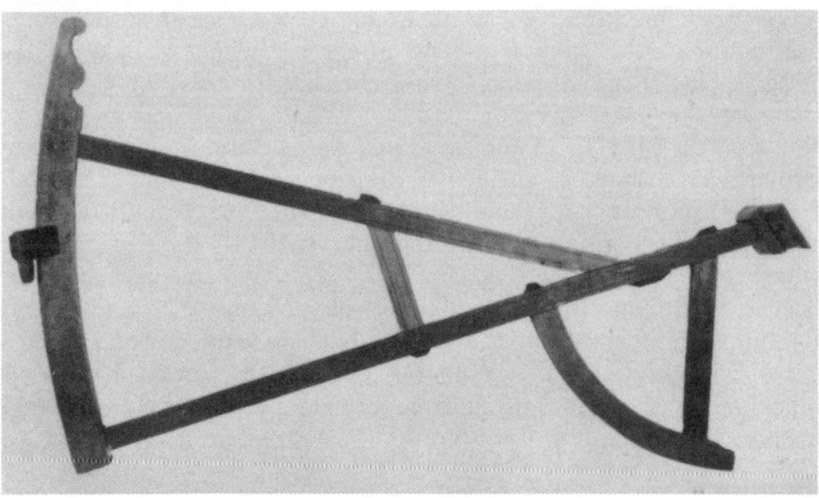

Figure 47.—Quadrant, showing signature of Thomas Greenough. Photo courtesy Connecticut Historical Society, Hartford.

Williams may have worked as an instrument maker in Marblehead before returning to his native Boston. According to Felt,[100] an instrument maker named William Williams at Marblehead advertised in the Salem newspapers in the early 1770's. However, in 1768 Williams was producing instruments from an address in King Street, Boston. (See figure 48.) An advertisement inserted by Williams appeared in the March 12, 1770, issue of *The Boston Gazette*. It was this same issue that reported the Boston Massacre. One of the victims was Williams' step-brother Samuel Maverick, the son of his stepfather Jotham Maverick by a first marriage.

In 1773 Williams married Joyce Shillcock, the daughter of his landlord. During the Revolutionary War, Williams saw active service as a private in Captain Mills' company, of Col. Jeduthan Baldwin's regiment of artificers, during the years 1777–1779. In 1780 he served in Captain Pattin's company of General Knox's artillery, which was stationed at West Point.[101]

With the conclusion of the war Williams returned to the craft of instrument-making in his shop, at No. 1 Long Wharf. In 1782 his wife, Joyce, inherited the property from her mother, the widow Hannah Shillcock, following the latter's death in that year. In the following May it is recorded that Williams purchased the warehouse and land on the north side of State Street from Benjamin Brown, a trader. By a separate deed, he and his wife released to Brown the warehouse and land which had been the property of his father-in-law in exchange for a clear title to one-half share of the store and land under it "which is next to the street called King Street." On February 7, 1784, he bought a share of the lower division at Long Wharf, No. 7, from Arnold Welles. On May 17 of the same year he succeeded in buying out Brown's half share of the lower division of Long Wharf at Nos. 1 and 7, and at the same time he deeded to Brown one-half share of No. 7 Long Wharf, together with all its dockage and wharfage. Finally, on January 20, 1785, Williams and his wife deeded to Brown all rights to land of No. 7 Long Wharf, reserving for himself his rights in the flats, wharfage, and dockage.

On March 23, 1787, Williams deeded to Joseph Helyer, a blockmaker, the store and land under same, and half the wharfage properly belonging to Lot No. 1. On October 20 of the same

---

[100] Felt, op. cit. (footnote 38), p. 173.

[101] "William Williams," in Thwing Catalogue, Massachusetts Historical Society.

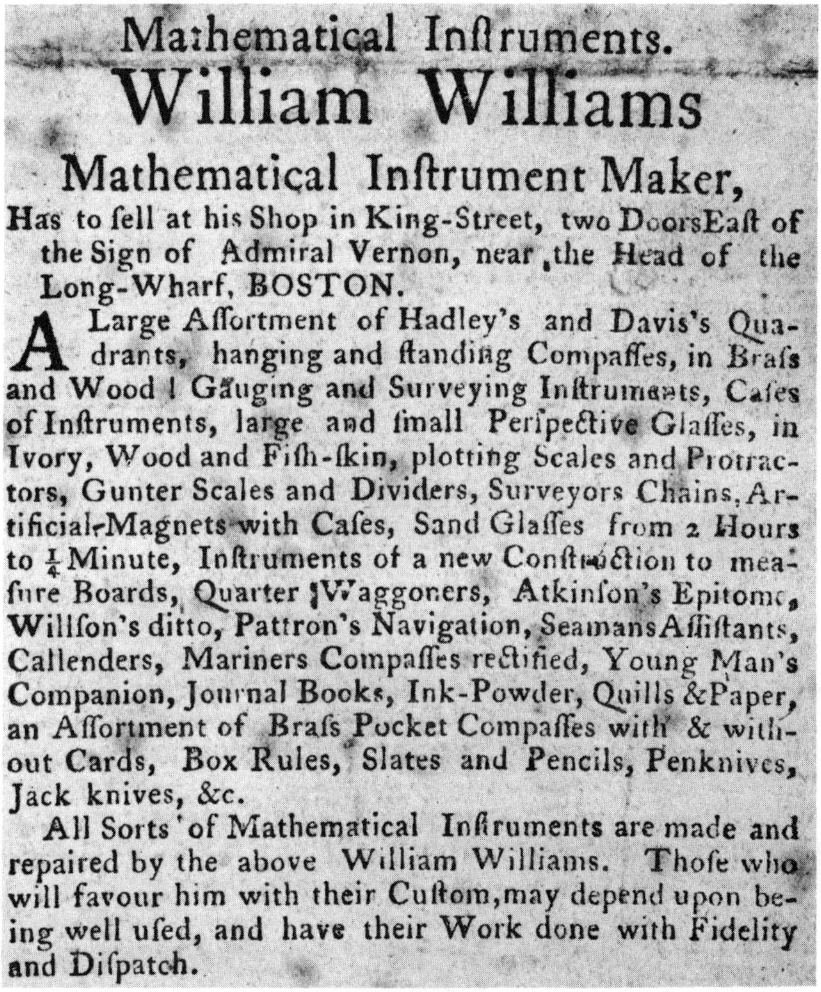

Mathematical Inſtruments.

William Williams

Mathematical Inſtrument Maker,

Has to ſell at his Shop in King-Street, two Doors Eaſt of the Sign of Admiral Vernon, near the Head of the Long-Wharf, BOSTON.

A Large Aſſortment of Hadley's and Davis's Quadrants, hanging and ſtanding Compaſſes, in Braſs and Wood ! Gauging and Surveying Inſtruments, Caſes of Inſtruments, large and ſmall Perſpective Glaſſes, in Ivory, Wood and Fiſh-ſkin, plotting Scales and Protractors, Gunter Scales and Dividers, Surveyors Chains, Artificial Magnets with Caſes, Sand Glaſſes from 2 Hours to ¼ Minute, Inſtruments of a new Conſtruction to meaſure Boards, Quarter Waggoners, Atkinſon's Epitome, Willſon's ditto, Pattron's Navigation, Seamans Aſſiſtants, Callenders, Mariners Compaſſes rectified, Young Man's Companion, Journal Books, Ink-Powder, Quills &Paper, an Aſſortment of Braſs Pocket Compaſſes with & without Cards, Box Rules, Slates and Pencils, Penknives, Jack knives, &c.

All Sorts of Mathematical Inſtruments are made and repaired by the above William Williams. Thoſe who will favour him with their Cuſtom, may depend upon being well uſed, and have their Work done with Fidelity and Diſpatch.

Figure 48.—Advertisement of William Williams in *The Boston Gazette*, March 12, 1770. Photo courtesy Harvard University Library.

year he sold to Brown a part or share of No. 7 Long Wharf, and on March 24, 1788, he purchased land with a wooden store at State Street and Long Wharf from Benjamin Brown. On June 26 he bought the land and store of Joseph Helyer on the north side of Long Wharf.

Williams engaged in only two more transactions before his death. On March 28, 1790, he mortgaged to Joseph Greene, a merchant, the land with wooden store at the head of Long Wharf on the northeast side of State Street; this mortgage was cancelled on May 29, 1793. On October 1, 1791, he deeded to Benjamin Brown a one-half share or 1/48th of all the dockage and wharfage of Long Wharf that appertained to one-half of Lot No. 1, which he had previously purchased from Welles as noted, as well as 1/48th of the proprietor's purchase of Gordon's lands and buildings adjoining the Wharf.

Williams died on January 15, 1792, at age 44. The administrator of his estate was a merchant named Abraham Quincy. By order of the Supreme Court, in order to settle his estate, Williams' store building at No. 1 Long Wharf was ordered sold at public auction. Although on the site of the Crown Coffee House, it was a new building erected in 1780 after the Coffee House had burned. The purchaser appears to have been John Osborn, a merchant, because on May 10, 1793, Quincy, Williams' administrator, deeded to Osborn the land with wooden store at Long Wharf on State Street.[102]

The only instrument made by Williams which appears to have survived is a Davis backstaff (fig. 49) marked "By Wm. Williams, King Street, Boston, for Malachi Allen, 1768"; this instrument is now in the collection of the Peabody Museum in Salem, Massachusetts. It is to be noted from this inscription that this instrument was an early example of Williams' work, produced at the age of 20, prior to the opening of his shop at the Crown Coffee House.

In 1770, when Williams opened his shop, the carved sign of "The Little Admiral" (fig. 37) was installed in front of the Crown Coffee House, and Williams' establishment was thereafter designated by this symbol.[103]

In his shop at No. 1 Long Wharf, Williams exercised his crafts of instrument- and clockmaking, and he made and sold a large

---

[102] Land record data from Thwing Catalogue, Massachusetts Historical Society.

[103] "Report of the Committee on the Rooms," *Proceedings of the Bostonian Society* (1917), no. 1, p. 16.

Figure 49.—Detail of wooden Davis quadrant inscribed "Made by William Williams in King Street Boston" for "Malachi Allen 1768." In collection of East India Marine Hall, Peabody Museum, Salem, Massachusetts.

assortment of instruments, as well as time glasses which measured from one quarter minute to two hours.

The name of Williams appears also in the Day Books of Paul Revere. Under date of April 16, 1792, there is the following entry:

> Mr. William Williams Dr
> To Engravg plate for hatt bills 0–18–0
> To 2 hund prints 0–6–0.

From June 24, 1792, to January 28, 1797, Revere entered 12 charges against Williams for 8,500 hat bills for the total amount of £ 14/15/0.[104]

Samuel Thaxter

Closely associated with the name of William Williams is that of another instrument maker of Boston, Samuel Thaxter (1769–1842). Thaxter was born in Hingham, Massachusetts, on December 13, 1769, the son of Samuel and Bathsheba (Lincoln) Thaxter. His father, who had been born in Hingham in 1744, was married on December 27, 1768, and he became the father of six

[104] BRIGHAM, op. cit. (footnote 32), p. 121.

children, of whom Samue was the eldest. Samuel Thaxter, Sr., was apparently a man of means, for he is listed as a "Gentleman" and a loyal subject of King George. He resided on North Street in Hingham, near Ship Street. He died on the island of Campobello at the age of 44 years on May 27, 1788.[105]

Samuel Thaxter, as well as several generations of his family before him, was born in the old Thomas Thaxter mansion that was built by the settler of that name in 1652. During the Revolution Samuel's father, Maj. Samuel Thaxter, concealed Tories from the Committee of Safety in a blind passage with a secret door in the old house. From there he smuggled them to Boston. At the massacre of Fort William, Major Thaxter was one of those captured by the Indians. While tied to a tree, he saw two French officers, and demanded whether this was the treatment they gave to commissioned officers. They allowed him to go free and he dragged himself to Fort Edward. Meanwhile, his comrades had reported him missing in action, and Dr. Gay preached his funeral sermon in Hingham shortly before Thaxter's return. The old Thaxter mansion was torn down in 1864.[106]

Young Samuel Thaxter moved from Hingham to Boston, where he is first heard of in 1792. On June 14, 1792, Thaxter married Polly Helyer, the niece of William Williams.

Within a month after the sale of Williams' property at public auction, Thaxter acquired the instrument-making business. Apparently the new owner of the premises required the business to move, and Thaxter established himself at No. 9 Butler's Row. A month after the Williams auction Thaxter announced his new location in an advertisement (fig. 50) in *The Columbia Centinel* of May 22, 1793.

Thaxter's new location was a wooden store structure, on the north side of Butler's Row that was owned by Andrew Hall and Eunice Fitch in 1798. It was in the rear of the north side of State Street, running from Merchants Row to the water.

By 1796 Thaxter had moved from this location to No. 49 State Street, on the north side opposite to Broad Street, a brick store owned by Joseph Lovering & Sons, tallow chandlers. He continued to do business at this address until 1815, when he moved to 27 State Street, on the opposite side of the street. The new loca-

[105] *History of Hingham* [Massachusetts], Hingham [n. d.], vol. 3, p. 236.

[106] Katherine M. Abbott, *Old Paths and Legends of New England* (New York: G. P. Putnam's Sons, 1909), pp. 341–342.

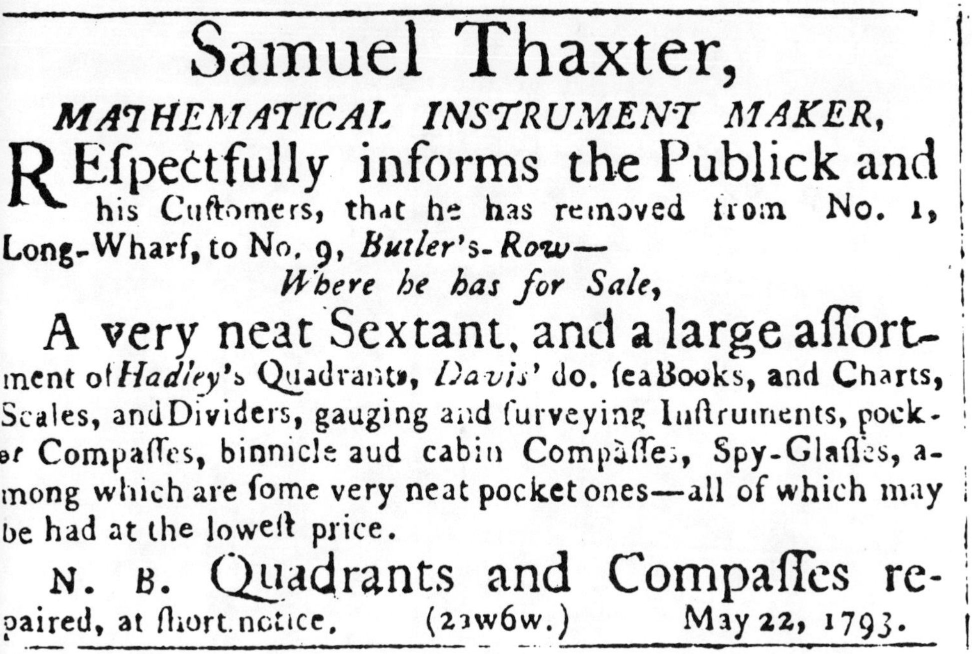
Samuel Thaxter,

MATHEMATICAL INSTRUMENT MAKER,

REſpectfully informs the Publick and his Cuſtomers, that he has removed from No. 1, Long-Wharf, to No. 9, *Butler's-Row*—

Where he has for Sale,

A very neat Sextant, and a large aſſortment of *Hadley's* Quadrants, *Davis'* do. ſeaBooks, and Charts, Scales, and Dividers, gauging and ſurveying Inſtruments, pocket Compaſſes, binnicle aud cabin Compaſſes, Spy-Glaſſes, among which are ſome very neat pocket ones—all of which may be had at the loweſt price.

N. B. Quadrants and Compaſſes repaired, at ſhort notice. (23w6w.) May 22, 1793.

Figure 50.—Advertisement of Samuel Thaxter in *The Columbia Centinel*, May 22, 1793. Photo courtesy Harvard University Library.

tion was in a brick dwelling, opposite Merchants Row, that was owned by Joseph Clough, a housewright.

In about 1825 Thaxter moved his business once more, to 125 State Street, the east corner of Broad Street. This building was occupied by Charles Stimpson, Jr., a stationer who was one of the publishers of the *Boston Annual Advertiser*, which was annexed to the Boston Directory of 1826. The building was owned by Jonathan Phillips, the first mayor of Boston. In the cellar of the building was a victualler named Augustus Adams.[107]

The dominating feature of Thaxter's shop from the time it was opened was the carved figure of "The Little Admiral," the trade sign first used by Williams.

The firm of Samuel Thaxter eventually became Samuel Thaxter & Son, and it continued with that name until past the middle of the 19th century. Samuel Thaxter died in April 1842 at the age of 72 years. The entry for the firm in the 1843 City Directory listed S. T. Cushing as the new owner. From the initials, it seems likely

[107] *Proceedings of the Bostonian Society* loc. cit. (footnote 103).

Figure 51.—19th-century trade card in collection of the Bostonian Society.

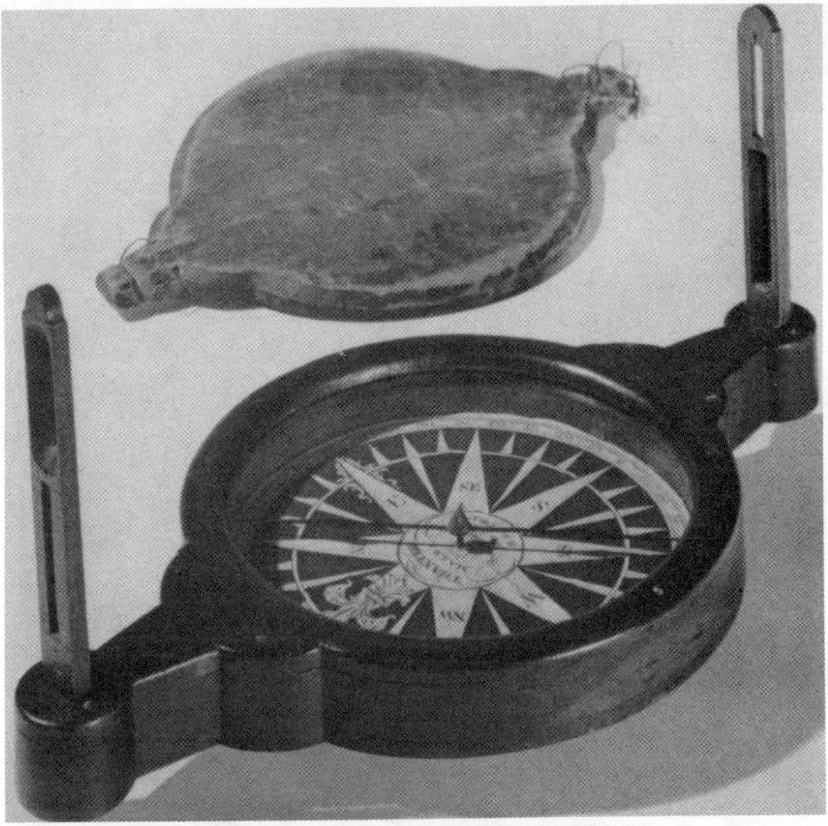

Figure 52.—Mahogany surveying compass made by Samuel Thaxter of Boston. Length, 13 in.; diameter, 7½ in. Wooden frame slides off to permit removal of glass and adjustment of needle. Sighting bars are of boxwood. In collection of the writer.

that his full name was Samuel Thaxter Cushing, and that he was the grandson of the original Samuel Thaxter. S. T. Cushing continued to be listed as the owner of the firm until 1899, when he was succeeded by A. T. Cushing, presumably a son of the former. The old store was finally demolished in 1901.[108] Comparison of a photograph of the building just before its demolition with a copy of Thaxter's trade card (fig. 51) of the mid-19th century shows that the building underwent little change in the period. The "Little Admiral" is barely visible in both views.

In 1796, shortly after his marriage, Thaxter made his home on

[108] Photograph and records in the collection of the Bostonian Society.

Figure 53.—Compass card from earlier form of wooden surveying compass made by Samuel Thaxter of Boston. From an instrument in the collection of the writer.

Fish Street (now North Street), but in 1800 he was living at 54 Middle Street (Hanover Street). By 1807 he had moved to a new home on Fleet Street. His last home address, at the time of his death, was 41 Pinckney Street.[109]

In the collection of the Massachusetts Historical Society there is a receipted bill (fig. 55) from Samuel Thaxter dated July 1, 1801, to Sam Brown, for touching up and repairing nine compasses for the French corvelle *Berceau*.

[109] Land records, Massachusetts Historical Society.

Figure 54.—Brass surveying compass made and sold by S. Thaxter & Son, Boston, in late 18th or early 19th century. Over-all length, 14 in.; diameter of dial, 6 in.; length of needle, 5⅛ in.; height of sighting bars, 6½ in. In collection of the writer.

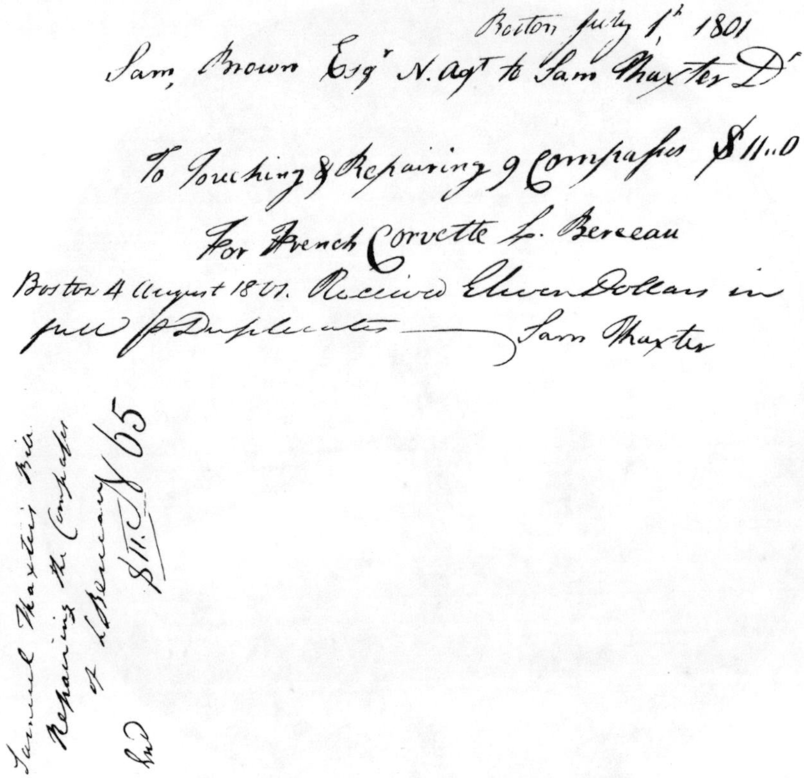
Boston July 1st 1801
Sam. Brown Esq. N. Agt to Sam Thaxter Dr
To Touching & Repairing 9 Compasses $11..0
For French Corvette L. Berceau
Boston 4 August 1801. Received Eleven Dollars in
full p Duplicate — Sam Thaxter

Samuel Thaxter's Bill
Repairing the Compasses
of L Berceau
$11.— 65

Figure 55.—Receipted bill from Samuel Thaxter to Sam Brown, Boston, August 4, 1801. In collection of Massachusetts Historical Society.

John Dupee

John Dupee of Boston apparently was another instrument maker of the pre-Revolutionary period actively engaged in producing wooden surveying compasses. Three wooden instruments with his compass card exist in private and public collections. The instruments are quite similar: the wood in each case is walnut or applewood, with an engraved paper mariner's compass card; a schooner at sea is figured within the central medallion, and inscribed within the riband enclosing it are the words "Made and Sold by JOHN DUPEE Ye North Side of Swing Bridge Boston New Eng." One of the instruments is owned by the South Natick [Massachusetts] Historical Society; a second example is in the collection of the Bostonian Society; and a third is owned by a private collector.

There is no record of a maker of scientific instruments or clocks by the name of Dupee, although the name John Dupee occurs in the city records of Boston during the early decades of the 18th century. An advertisement in the February 9, 1761, issue of *The Boston Gazette* states that

> ISAAC DUPEE, Carver, Advertises his Customers and others, that since the late Fire (on Dock Square) he has opened a shop the North side of the Swing-Bridge, opposite to *Thomas Tyler's*, Esq.; where Business will be carried on as usual with Fidelity and Dispatch.

The natural assumption would be that the three instruments were produced in Isaac Dupee's shop after 1761, perhaps by the carver's son. The use of an engraved compass card indicates that the instruments were not unique, and that a number of others were produced or contemplated. On the other hand, it is likely that the maker produced other types of instruments utilizing such a card, such as mariner's compasses.

Jere Clough

Another instrument maker, presumably of Boston, is Jere Clough. The only instrument bearing his name known at present is a surveying compass (fig. 56), made of wood, in the Streeter Collection of Weights and Measures at Yale University. Clough's name does not appear on any of the lists of instrument makers or clockmakers, yet it is a name that is fairly prevalent in Boston. In 1741, for instance, one Joseph Clough of Boston was a maker of bellows.

Figure 56.—Wooden instrument made by Jere Clough. In Streeter Collection of Weights and Measures, Yale University.

Figure 57.—Wooden surveying compass made by Andrew Newell (1749–1798) of Boston. It is made of mahogany, is 11½ in. long, and has a diameter of 5 in. The engraved compass card is signed by Nathaniel Hurd, goldsmith, silversmith, and engraver of Boston. In collection of Yale University Art Gallery.

He produced bellows of all types—for furnaces, refiners, blacksmiths, braziers, and goldsmiths.[110]

Andrew Newell

An instrument of considerable significance is another wooden surveyor's compass, in the collection of the Yale University Art Gallery. This compass (fig. 57) is made of rich brown San Domingo mahogany with sighting bars of boxwood. A mariner's card, set into the opening with a metal vernier scale, is in the usual form of the mariner's compass card of the 18th century; it is executed as a line engraving. A ship and the Boston harbor lighthouse are featured in the central medallion. On a riband encircling the medallion is the inscription "Made by ANDW. NEWELL East End of the MARKET BOSTON," Engraved in script at the southern tip of the star is the signature "N. Hurd Sct."

Relatively little is known about Andrew Newell (1749–1798) except that he was a maker of mathematical instruments. An entry in the first Boston directory, in 1789, listed "Andrew Newell, instrument maker, 61 State Street." The directory of 1796 mentioned Newell as having a shop on the "East side of the Market," the address that appears on the surveying compass.

[110] GEORGE FRANCIS DOW, *The Arts and Crafts in New England 1704–1775* (Topsfield, Mass.: The Wayside Press, 1927), p. 256.

Two years later the Boston directory listed Andrew Newell and Son, and in 1800 the listing included only the name of Joseph Newell, who may have been the son. Another mathematical instrument maker named Charles Newell may have been another son of Andrew Newell; his name does not appear in the city Directory until in the 19th century. An instrument with the signature "Newell & Son, Makers, East End of Faneuil Hall, Boston" is in the collection of the Bostonian Society.

An important feature of the Newell instrument is the fact that the engraver of the compass card was Nathaniel Hurd (1729–1777), the peer of goldsmiths and engravers of the colonial period. This compass card is a previously unrecorded example of Hurd's work, and constitutes a work of art, making the compass a historic scientific instrument.[111] The compass was presented to the Yale University Art Gallery by a Yale alumnus, Mr. Henry G. Schiff of New York City. No other examples have thus far been found.

Aaron Breed

Aaron Breed (1791–1861) is a relatively unknown maker of mathematical instruments who worked in Boston into the 19th century. He specialized in nautical, mathematical and optical instruments, with an address at 173 Broad Street, and another at No. 2 Rowe's Wharf, "At the Sign of the Quadrant." Breed made surveying instruments in brass and in wood. A brass instrument is in the Henry Ford Museum, and a wooden instrument is in the collection of Old Sturbridge Village. The latter is fashioned from walnut with an engraved compass card inscribed "Aaron Breed Boston."

Charles Thacher

The name of Charles Thacher appears on the compass card of a wooden surveying compass (fig. 58) in the collection of the Mariners' Museum, Norfolk, Virginia. No record of this maker has been found, but the engraved compass card indicates that he probably worked in New England.

---

[111] John M. Phillips, "An Unrecorded Engraving by Nathaniel Hurd," *Bulletin of the Associates in Fine Arts at Yale University* (June 1936), vol. 7, no. 2, pp. 26–27.

Figure 58.—Wooden surveying compass made by Charles Thacher. It is made of cherry or maple; sighting bars are of oak. Over-all length, 13⅝ in. Photos courtesy Mariners Museum, Newport News, Virginia.

Benjamin King Hagger

Benjamin King Hagger (c. 1769–1834) was the scion of two well-known families of instrument makers in New England, so it is not surprising that he worked in the same craft.

It is believed that Hagger was born in Newport, Rhode Island, about 1769, the son of William Guyse Hagger and of a sister of Benjamin King. Although his father made instruments—at first in partnership with Benjamin King, and then working alone—in Newport at least as late as 1776, the family appears to have moved after the Revolution. William Guyse Hagger's name did not appear in the 1790 census of Newport, and it is presumed that he moved with his family to Boston.

Benjamin King Hagger was listed in the first city directory of Boston in 1789 as "a mathematical instrument maker" with an address on Ann Street; he was only 20 years of age at this time.

On November 10, 1793, Benjamin King Hagger, "mathematical instrument maker," purchased land with buildings on Prince Street near Snow Hill Street from one Peter Greene. Two years later, on December 1, 1795, Hagger, now listed simply as a "merchant," purchased a brick house, a wooden house, and a shed with land from William Ballard, a tailor of Framingham and an heir of Samuel Ballard. The property was located on the east side of North Street, south of Mill Creek. At the time of purchase, Hagger mortgaged the property to Ballard, and also mortgaged to him the house and land previously purchased from Greene.

Hagger was listed as a ship chandler in the following year when on March 24, 1796, he deeded part of his land on Prince Street to William and George Hillman, minors.

On June 22, 1796, three months later, Hagger, now listed as "mathematical instrument maker, and ship-chandler" deeded to a mariner named Thomas Wallis a house and land that formed part of his original purchase near Copp's Hill from Peter Greene. Then on July 21, 1796, he purchased from William Ballard all his right to the brick house and land on North Street (Ann Street), at the same time mortgaging the property to William Ballard, Jr., of Framingham. This mortgage was cancelled on April 11, 1798.[112]

These negotiations took place before marriage. A report of the Record Commissioners of Boston, states that "William King Hagger of Boston and Mehitable Ballard of Framingham were

[112] Land records on Benjamin King Hagger listed in Thwing Catalogue, Massachusetts Historical Society.

married October 6, 1796." The entry appears to be in error because the marriage intentions had read "Benjamin King Hagger." It is presumed that Mehitable was the daughter of William Ballard, the tailor of Framingham, from whom Hagger had bought his house on Ann Street, south of Mill Creek.[113]

Benjamin King Hagger is listed in the city directory of Boston for 1798 as a "mathematical instrument maker" on Ann Street. This, however, is the last listing for his name in Boston, as his name does not appear in the 1803 or subsequent directories.

Shortly after 1798 Hagger appears to have left Boston together with his wife, and it is probable that he established himself as an instrument maker in another Massachusetts community, at present unknown. In about 1816 Hagger moved with his family to Baltimore and continued his instrument-making business.

The records of the 1850 Federal census of Baltimore indicate that two of Hagger's sons, John W. and William G. Hagger, had been born in 1800 and 1806 respectively, in Massachusetts, presumably in the community to which Hagger had moved from Boston before moving once more to Baltimore.

According to Matchett's Baltimore directory for 1824, Hagger was a "mathematical and optical instrument maker" with a shop at 57 South Street. His advertisement in the directory stated that he

> Respectfully acquaints his fellow citizens that he executes all orders in the line of his business with punctuality and confidently professes to give satisfaction to his employers, from the experience of a regular apprenticeship and 37 years practice.

This indicates that Hagger completed his apprenticeship in 1787, when he was 18, and since then had been established in his own business or had worked for another as a journeyman instrument maker. His first advertisement in the Boston directory appeared in 1789, wherein his shop was listed as being on Ann Street.

Hagger died in Baltimore on November 8, 1834, at the age of 65, after a residence of 18 years in that city.[114]

Thus far only one instrument by Hagger has been found—a wooden surveying instrument or semicircumferentor (fig. 59). It is in the possession of the writer.

---

[113] Marriage Document no. 101, Report of the Record Commissioners of Boston, p. 298.

[114] *The Baltimore American and Commercial Advertiser*, November 9, 1834.

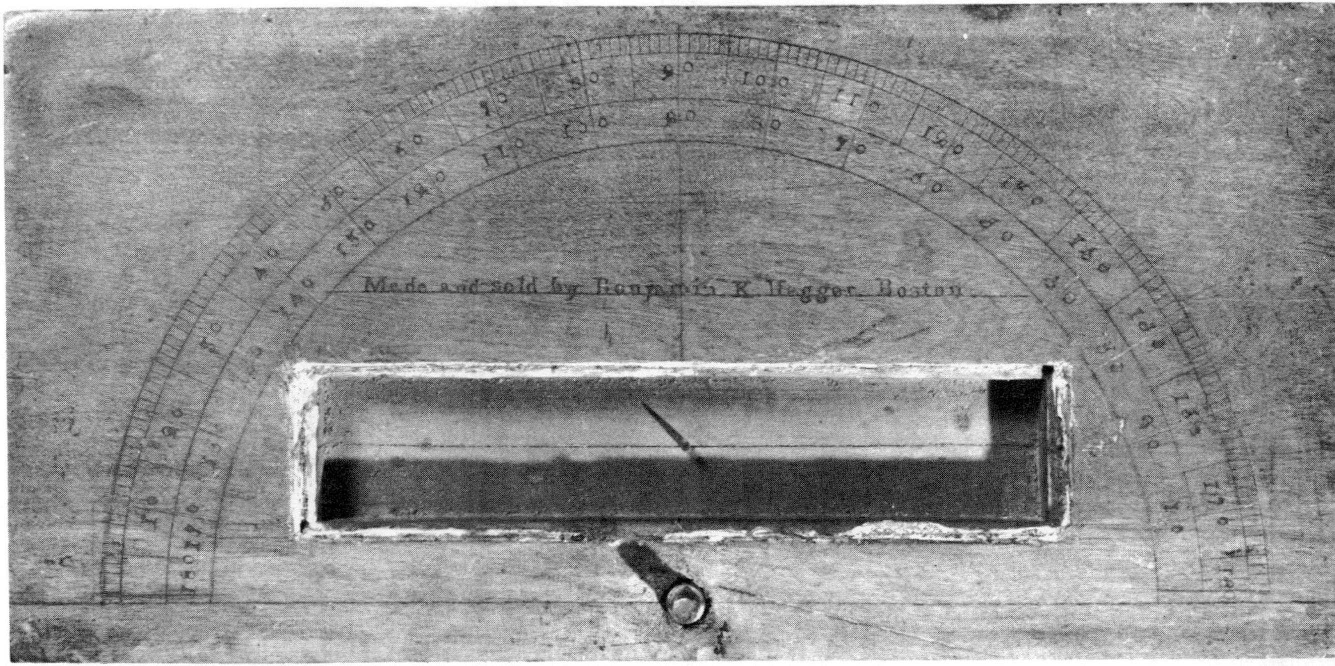

Figure 59.—Wooden graphometer made by Benjamin King Hagger (c.1769–1834) of Boston and Baltimore. Made of yellow birch, with the name and gradations and lines incised into the wood by means of tiny punches, and filled. Trough compass; sighting bars mounted on a swivelling brass bar; collapsible tripod made of maple. In collection of the writer.

Benjamin Warren

Production of wooden surveying compasses was not limited to Boston. Another instrument maker who produced them was Benjamin Warren (c. 1740–?) of Plymouth, Massachusetts. The name of Benjamin Warren was a fairly common one in Plymouth, being a name handed down in the family from father to son for at least five generations before 1800. The first Benjamin Warren at Plymouth was married in 1697, and his son Benjamin (2) was born in

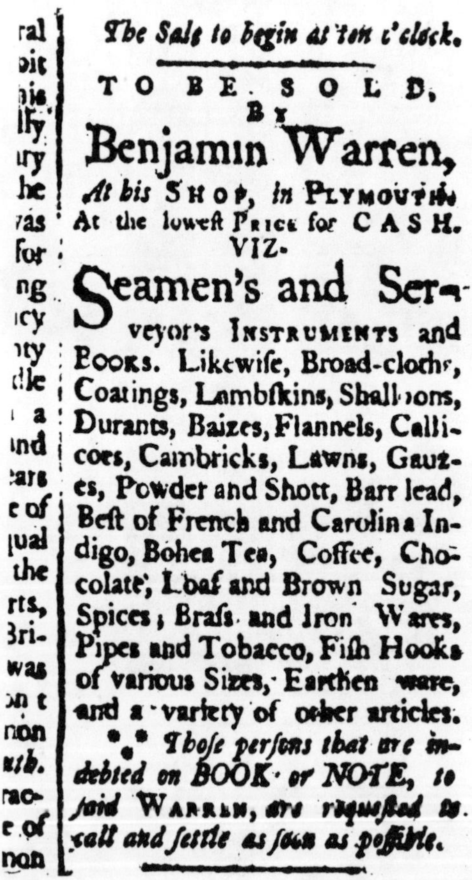

Figure 60.—An advertisement of Benjamin Warren in *The Plymouth Journal & Massachusetts Advertiser*. Photos courtesy The American Antiquarian Society, Worcester, Massachusetts.

1698. Benjamin (2) was married in due course, and his son Benjamin (3) was born in 1740. The third Benjamin was the father of Benjamin (4), who was born in 1766. In 1789 Benjamin (4) married Sarah Young, the daughter of Daniel Young, and their son Benjamin (5) was born in 1792. The Benjamin Warren who operated the shop in Plymouth probably was Benjamin Warren (3), who was then about 45 years of age.[115]

A search of *The Plymouth Journal & Massachusetts Advertiser* has revealed several advertisements and notices (fig. 60) about Benjamin Warren from which some information can be derived about the man and his business during this period. The first known notice dated March 19, 1785, probably is the most important one. Later in the same year, on August 16, 1785, Warren published the following notice:

> WHEREAS on Friday Morning of the 5th inst. eloped from the House of the subscriber, *Inholder* in Plymouth, JOHN MOREY, of NORTON, of tall stature, & round shoulder'd. Had on when he absconded, a shabby claret coloured coat, adorned with patches, and a pair of dirty smoak'd coloured breeches; without knee-buckles; and an old flopped hatt, defaced with grease.
>
> As he appeared to be an enterprising genius, without abilities, politeness or honesty, and went off in an abrupt and clandestine manner; a reward of *Sixpence* will be paid, to any person or persons, who will persuade or induce the said Morey to make his appearance once more to the subscriber.

It is obvious that Warren was not considerably concerned about the return of John Morey, for the reward offered was scarcely conducive to obtain the public's cooperation. Warren's first ventures with public sales must have been successful, for early in the next year, in the issue of January 3, 1786, he announced that

> *Benjamin Warren*,
>
> PROPOSES to open a convenient AUCTION-ROOM, over the Shop he now trades in, next week. Any Gentlemen that will furnish him with goods of any kind for Public or Private sale, on Commission, shall be served with fidelity, and the smallest favours in that way gratefully acknowledged.

The next notice of the auction-room appeared on February 21, 1786, when the newspaper advertised that

---

[115] Silvio A. Bedini, "A Compass Card by Paul Revere (?)", *Yale Library Gazette* (July 1962), no. 2. pp. 36–38; William T. Davis, *Ancient Landmarks of Plymouth* (Boston: A. Williams & Co., 1883).

To-morrow
will be SOLD, by
Public Vendue,
At
WARREN'S
Auction Room,
A VARIETY of articles, *viz.* Nails, Bar Lead, Glass
Pewter, Buttons, Buckles, Chairs, Stands, &c, &c, &c.
*** The SALE to begin at 10 o'Clock, A.M.

No other notices of public sales appeared in the *Journal* for the next several months. The last notice of this period was another announcement of a sale, which was published in the issue of May 30, 1786:

Publick Vendue,

At WARREN's Auction Room, in PLYMOUTH: at Ten o'clock this morning. WILL be Sold, a quantity of bar lead, boxes of glass, 6 x 8. English Shovels and Tongs, bridle-Bits, and a variety of other articles of Hard-Ware. Also, a few Anvils at private sale.

Only one instrument signed by Warren is known to survive; it is a wooden surveying compass (fig. 61) in the Streeter Collection

Figure 61.—Wooden surveying compass made by Benjamin Warren (c.1740–c.1800) of Plymouth, Massachusetts, and detail of the compass card. The compass, made of cherry wood, is 12 in. long and has a diameter of 6 in. In Streeter Collection of Weights and Measures, Yale University.

Figure 62.—Detail of card, Warren surveying compass shown in figure 61.

of Weights and Measures at Yale University. The instrument, which appears to have been made from walnut, has a compass card with the following inscription around the central medallion: "Made and sold by BENJAMIN WARREN Plymouth New Eng^d."

The medallion (fig. 62) encloses a harbor scene with a brigantine of the 1740 period off a promontory on which is prominently situated a lighthouse with a smaller building partly visible at the left. The lighthouse is unusual in construction in that it features twin towers rising from a large rectangular wooden building.

As far as can be determined from available records, the only lighthouse in America of this period having such construction was the noted Gurnet Light, which was built at the tip of Duxbury Beach in Plymouth Bay in 1768. D. Alan Stevenson [116] relates that the Governor's Council of Massachusetts, when it decided in 1768

---

[116] D. ALAN STEVENSON, *The World's Lighthouses before 1820* (London: Oxford University Press, 1959), p. 179.

to erect the Gurnet Lighthouse at Plymouth, adopted a novel plan to distinguish it from other American lighthouses. "This consisted of double lights set horizontally in the same structure. A timber house built at a cost of £660, 30′ long and 20′ high, had a lanthorn at each end to contain two four-wick lamps.

"In 1802 fire destroyed the house but the merchants of the town promptly subscribed to replace it by temporary lights, as the Government had no immediate funds at its disposal. An Act of Congress of 1802 allotted $2500 for building another set of twin lights and reimbursing the merchants for their expenditure.

"Though the idea of twin lights at Plymouth seemed an excellent distinction from a single navigation light shown at Barnstable harbor in the vicinity, they proved not entirely advantageous and a sea captain blamed them for causing his shipwreck. He had seen the light from only one tower and identified it with confidence as the Barnstable light; apparently, from a particular direction one tower hid the other. But local prejudice in favor of retaining the twin lights as a distinction prevailed until 1924 when, at last, opposition ceased to the recommendation which the Lighthouse Board expressed frequently that a single light would be preferable."

It seems quite likely that the compass card bears one of the very few surviving contemporary representations of the first Gurnet Light in Plymouth Bay. A search of the archives of the historical societies in Plymouth, Boston, and Worcester and the files of the U.S. National Archives has failed to reveal any illustration of this famous lighthouse.

Quite by coincidence, the name of Benjamin Warren was discovered among the entries of the day books of Paul Revere, the famous patriot, silversmith, and engraver. The entry [117] (fig. 63) appears as follows:

1786 March 13. Benjm Warren Dr. Plimouth
To printing one hundred Compass Cards 0–18–0.

Whether the compass card on the Warren instrument was produced by Revere is difficult to determine. Authorities on Revere's engravings agree that it could have been engraved by Revere but are unable to state it positively. It has been suggested that the entry in Revere's day book indicates that he merely printed the compass cards for Warren and that he did not engrave a plate. The charge for the work bears out this supposition; and furthermore, Revere's bills seemed to make a definite distinction

[117] Paul Revere, *Day Books*, MS., Massachusetts Historical Society.

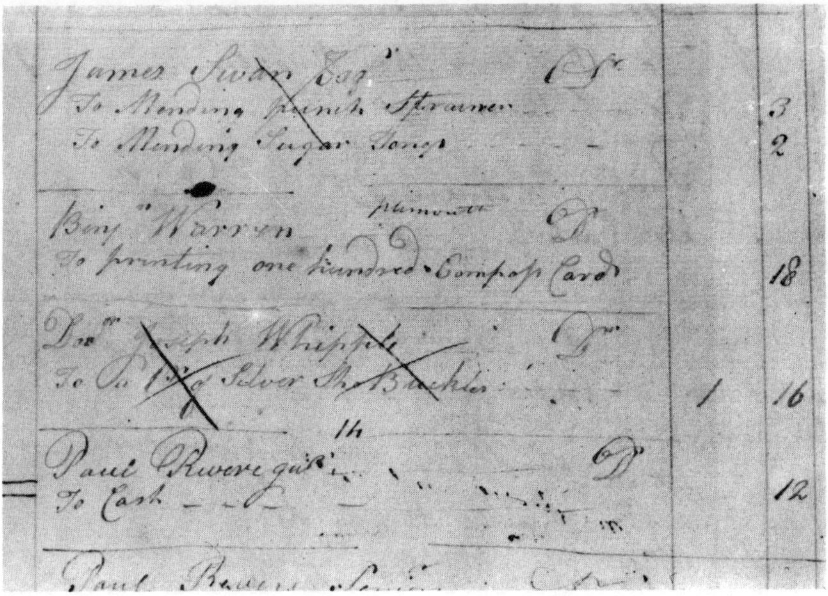

James Swan Esq. Cr
To Mending Punch Strainer — 3
To Mending Sugar Tongs — 2

Benj. Warren Plymouth Dr
To printing one hundred Compass Cards — 18

Doctr Joseph Whipple Dr
To a Pr of Silver Sh Buckles — 1 16

Paul Revere Junr Dr
To Cash — 12

Figure 63.—Page from the "day books" of Paul Revere with entry for the printing of compass cards for Benjamin Warren of Plymouth. In collection of Massachusetts Historical Society, Boston.

between the engraving of plates and actual prints. Whether or not Revere was responsible for making the original engraving remains to be determined, but it is very probable that he printed the compass card of the instrument in the Streeter Collection of Weights and Measures at Yale.

Daniel Burnap

One of the best known and most respected names among Connecticut clockmakers is that of Daniel Burnap (1759–1838) of East Windsor. Burnap was born in Coventry in 1759 and served an apprenticeship with Thomas Harland, clockmaker of Norwich. In about 1780 Burnap opened his own establishment, where he combined the crafts of clockmaking, cabinetmaking, and engraving of brass, in all of which he was greatly skilled. One of his apprentices was Eli Terry, who later achieved fame in the craft in his own right.[118]

Burnap's business included clients in Windsor, Hartford, and

[118] Hoopes, op. cit. (footnote 50), pp. 7–8.

Coventry, as well as some of the leading merchants and cabinet-makers of the nearby cities and towns. Although clockmaking was the primary business in which Burnap engaged, he also had a large trade for his surveying instruments, silver spoons, gold beads, harness and saddlery hardware, and shoe buckles.

Burnap prospered, and in about 1800 he moved back to his native town, Coventry. There he purchased a large farm and erected a shop and a sawmill, and in due course became the leading citizen of the community. He died in 1838, leaving a valuable technological record in the completeness of his journals and account books. A study of the entries of his day books and ledgers (see fig. 64) reveals that Burnap did a substantial amount of business in surveying compasses, chains, and protractors. Among his shop equipment after his death there was found an unfinished protractor, but no examples of his instruments are known except for a compass dial, inscribed with his name, that was discovered recently in the collection of a midwestern historical society.[119]

It is significant to note that Burnap made instruments of varying quality. For instance, he charged three different prices for his surveyor's compasses. The highest-priced compasses cost £6; they were made of brass, and were of the more elaborate conventional type used by surveyors. A few examples that appeared in his records cost £4; these also were made of brass, but probably were of a simpler form. Several entries list surveying compasses priced at £2 and £2/8. One of these was made for Capt. Solomon Dewie (1750–1813) in September 1790 for £2/8. At the same time, Burnap charged him £0/1/6 for touching the needle of another compass.[120] The entries in Burnap's account books do not state that these inexpensive compasses were constructed of wood, but it seems to be sufficiently conclusive that they were.

Gurdon Huntington

Gurdon Huntington (1763–1804) was not primarily a maker of scientific instruments, but he was established as a goldsmith and clockmaker. He was born in Windham, Connecticut, on

---

[119] Information from Mr. C. E. Smart, of W. & L. E. Gurley, Troy, New York.

[120] Penrose R. Hoopes, *Shop Records of Daniel Burnap, Clockmaker*, (Hartford: Connecticut Historical Society, 1958), pp. 63–66.

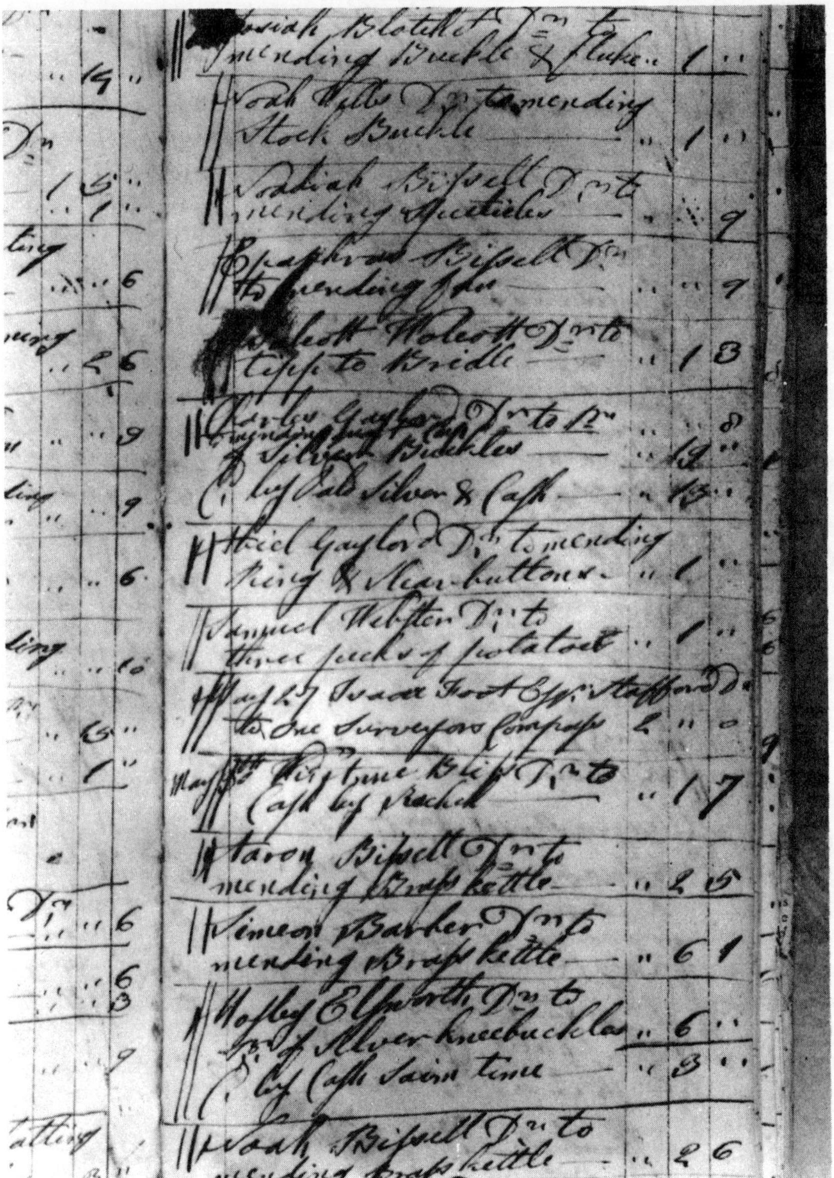

Figure 64.—Entry in the manuscript ledgers of Daniel Burnap (1759–1838) of East Windsor and Coventry, Connecticut, for sale of surveying compass in 1790. Reproduced from the Burnap shop records in the collection of Connecticut Historical Society, Hartford.

April 30, 1763, the son of Hezekiah and Submit (Murdock) Huntington.[121]

The Huntington family was one of the most important in Connecticut colonial history. Gurdon's father, Hezekiah, was in service during the Revolutionary War, going to Boston as a major with the first troops raised in Connecticut. When in Boston he witnessed the miserable condition of the arms then in the hands of the soldiers. Major Huntington went immediately to Philadelphia, where Congress was in session, and proposed to the Congress that he would return to his home in Windham and that there he would open a manufactory for repairing muskets and other arms. He claimed to have been the first man to have made a gun in the Colonies.

Gurdon was too young to have served in the Revolution, but he undoubtedly worked in his father's gun manufactory as a boy. In due course he learned the trades of goldsmith and clockmaker and established his own shop in Windham, which, according to an advertisement (fig. 65) in *The Connecticut Gazette* of June 11, 1784, was "a few rods north of Major Ebenezer Backus' store."

On Christmas Day, 1785, Gurdon was married in New London to Temperance Williams of Groton. In 1789 their first child, Marvin, was born, and in October of the same year the Huntingtons moved from Windham to Walpole, New Hampshire. No reason can be found for the move, other than the possibility that Gurdon might have anticipated greater opportunity in the new community. There he applied himself to his trade as goldsmith and clockmaker, but apparently he was not very successful. His family grew, and by the time of his death there were eight children. Possibly in an effort to supplement his income, Huntington served as postmaster of the community. In about 1797, seven or eight years after he had moved to Walpole, his father and mother joined him there, and it is believed that Major Hezekiah may have worked as a gunsmith during that period. Eventually the senior Huntington returned to Windham, Connecticut, where he died in 1807.[122]

Meanwhile Gurdon Huntington struggled on until his death on July 26, 1804. He died insolvent, which created a considerable problem in view of the large family he left behind him. Huntington's estate was administered by Asa Sibley, a clockmaker in

[121] Hoopes, op. cit. (footnote 18), pp. 92–93.

[122] *Memoirs of the Huntington Family Association* (Hartford, Conn., 1915), Index no. 1.3.4.4.2.4.

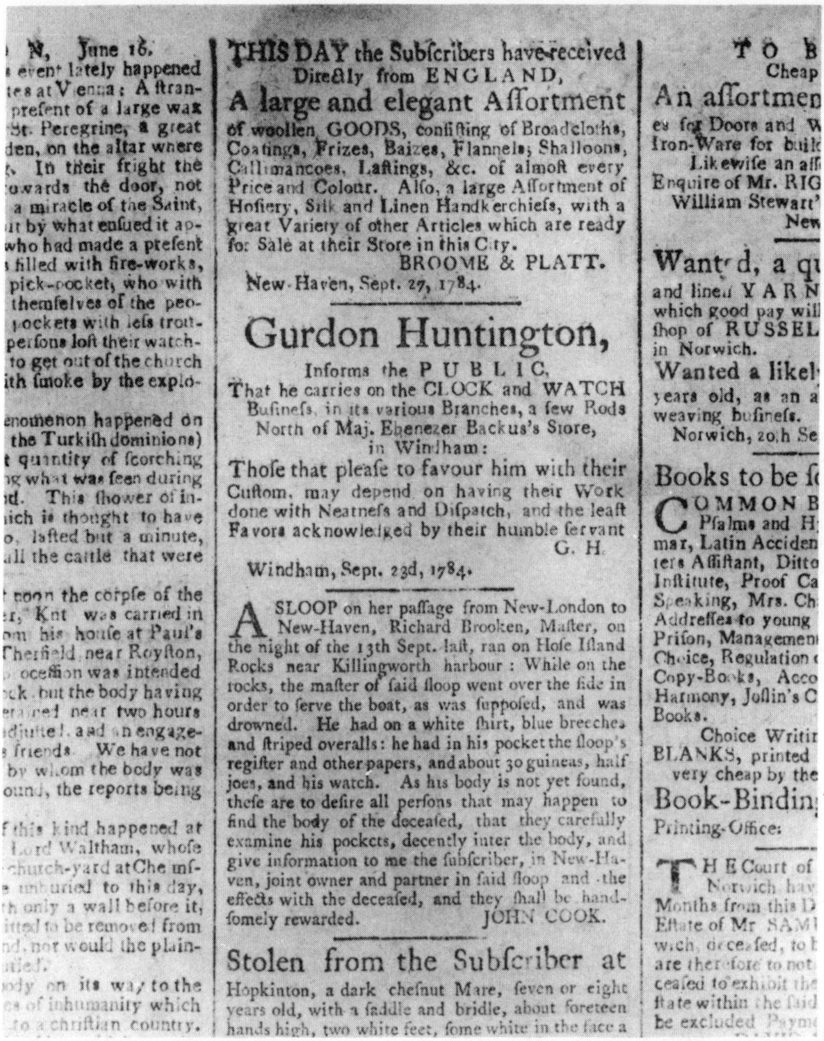

THIS DAY the Subscribers have received
Directly from ENGLAND,
A large and elegant Assortment
of woollen GOODS, consisting of Broadcloths, Coatings, Frizes, Baizes, Flannels, Shalloons, Callimancoes, Lastings, &c. of almost every Price and Colour. Also, a large Assortment of Hosiery, Silk and Linen Handkerchiefs, with a great Variety of other Articles which are ready for Sale at their Store in this City.
BROOME & PLATT.
New-Haven, Sept. 27, 1784.

Gurdon Huntington,
Informs the PUBLIC,
That he carries on the CLOCK and WATCH Business, in its various Branches, a few Rods North of Maj. Ebenezer Backus's Store, in Windham:
Those that please to favour him with their Custom, may depend on having their Work done with Neatness and Dispatch, and the least Favors acknowledged by their humble servant
G. H.
Windham, Sept. 23d, 1784.

A SLOOP on her passage from New-London to New-Haven, Richard Brooken, Master, on the night of the 13th Sept. last, ran on Hose Island Rocks near Killingworth harbour: While on the rocks, the master of said sloop went over the side in order to serve the boat, as was supposed, and was drowned. He had on a white shirt, blue breeches and striped overalls: he had in his pocket the sloop's register and other papers, and about 30 guineas, half joes, and his watch. As his body is not yet found, these are to desire all persons that may happen to find the body of the deceased, that they carefully examine his pockets, decently inter the body, and give information to me the subscriber, in New-Haven, joint owner and partner in said sloop and the effects with the deceased, and they shall be handsomely rewarded.
JOHN COOK.

Stolen from the Subscriber at
Hopkinton, a dark chesnut Mare, seven or eight years old, with a saddle and bridle, about foreteen hands high, two white feet, some white in the face a

Figure 65.—Advertisement of Gurdon Huntington (1763–1804) in *The Connecticut Gazette*, June 11, 1784. In collection of Connecticut Historical Society, Hartford.

Walpole. Sibley had moved to Walpole from his home in Woodstock, Connecticut, in the 1790's and he remained there until 1808, when he again returned to Woodstock. Gurdon Huntington's widow removed to Bloomfield, Ohio, with her children, and she died there on May 25, 1823. Most of her children settled

Figure 66.—Views of wooden surveying compass made by Gurdon Huntington, clockmaker in Walpole, New Hampshire, between 1789–1804. Made of cherry with folding brass sighting bars, the instrument is 14 in. long and 5½ in. wide. In collection of the writer.

in Bloomfield, but several of them moved to New Hartford, New York.

Several examples of Huntington's clocks are known to exist in private collections in the United States. However, only one example of his scientific instruments appears to have survived. This is a surveying compass (fig. 66) made of wood, with brass sighting bars and a painted dial under glass with a steel needle. The dial is inscribed "G. HUNTINGTON/WALPOLE." The instrument, which is in the collection of the writer, is made of cherry wood, with a riveted ball-and-socket joint of brass for insertion on a tripod.

Jedidiah Baldwin

Jedidiah Baldwin (fl. 1790's) was another early New England clock and instrument maker, but little is known of his early life. He was a brother of Jabes Baldwin (c. 1777–1829), who worked as a clockmaker in Salem and Boston after serving an apprenticeship with Thomas Harland in Norwich, Connecticut.

Jedidiah Baldwin also served an apprenticeship with Harland. In 1791 he was working in Northampton, Massachusetts, as a member of the firm of Stiles and Baldwin, and from 1792 to 1794 he was a member of the firm of Stiles and Storrs, in partnership with Nathan Storrs.[123] In about 1794 Baldwin moved to Hanover,

[123] PALMER, op. cit. (footnote 34), p. 143.

New Hampshire, where he became the local postmaster, and where Dartmouth College records his death.

Only one existing instrument is known to have been made by Baldwin; it is a wooden surveying compass with a brass dial having two scales, one for degrees and one for eight divisions per 90°. The dial is inscribed "JED BALDWIN/HANOVER." According to its present owner, Mr. Worth Shampeny of Rochester, Vermont, the compass was used for surveying in Vermont during the early 1800's.

Another Jedidiah Baldwin worked as a clockmaker in Morrisville, New York, from 1818–1820 and then in Fairfield, New York; he appears also in the city directory of Rochester, New York, as a clockmaker during the years 1834–1844. He may have been a son or grandson of the first Jedidiah, or a nephew.

Thomas Salter Bowles

Thomas Salter Bowles (c. 1765–?) is another elusive New England instrument maker about whom little information is available. He is believed to have been the son of Deacon Samuel and Hannah (Salter) Bowles, born in Portsmouth, New Hampshire, probably between 1765 and 1770. His father was born in 1739; his mother, who was the daughter of Captain Titus Salter, was born in 1748 and died in 1831.[124] Deacon Bowles was clerk of the Brick Market in Portsmouth from 1801 to the time of his death, November 3, 1802. There is a minimum of information available from church and city records in the community, but it is believed that he was a member of one of the offshoots of the established Puritan Church, and hence he would not appear in its records. He kept the lower school in the Brick School House on State Street for a number of years.

It is believed that the Bowles family first came to Portsmouth during the few years immediately before the beginning of the Revolutionary War. It is known that a Thomas Bowles and a Samuel Bowles both signed the Association Test on August 14, 1776, promising to oppose the hostile proceedings of the British fleets and armies. Furthermore, one of the principal taxpayers in Portsmouth in 1770 was a firm named Griffith and Bowles, which paid £17 in taxes in 1770. The name of the Bowles who formed part of this firm is not known, but it was either Samuel or the first

---

[124] Correspondence with Mr. Ray Brighton, Portsmouth, N. H.

Thomas Bowles. The other partner was Nathaniel S. Griffith, a watchmaker. It is possible that a tradition of instrument making existed in the Bowles family even then.[125]

On file in the office of the City Clerk in Portsmouth are two certificates of marriage made out by Thomas Salter Bowles. The first is for his marriage to Hannah Ham, a ceremony performed on September 21, 1809, by Joseph Walton, one of the pastors of a church dissenting from the Puritan regime. Hannah was the daughter of William Ham, a brother of Supply Ham (1788–1862), a noted local clockmaker. Bowles may have served an apprenticeship in that shop before he married Hannah. Two other members of the Ham family—George Ham and Henry H. Ham—worked as watchmakers in Portsmouth in the same period.

A search of the cemeteries has indicated that Hannah Ham Bowles died in 1811, age 20. She is buried with her infant son in North Cemetery.[126]

Thomas Bowles's second marriage certificate in Portsmouth is for his marriage on September 29, 1813—two years after Hannah's death—to Abiah Emerly Bradley of Haverhill, Massachusetts.

Little is known about the work of Bowles as an instrument maker except through a few of his instruments. He is listed in the first Portsmouth directory, of 1821, as a "mathematical instrument maker" with a place of business on Daniel Street; his home was given as Austin Street in Portsmouth. He did not appear in the city's directories of 1827 and 1834. It is assumed that he may have left Portsmouth in the interim, possibly to settle in his wife's home town of Haverhill.

Three instruments signed by Bowles have survived, and all show signs of considerable wear. They are surveying compasses made of walnut, having maple sighting bars and a silvered brass vernier set under the glass. Two examples, one in the Streeter Collection of Weights and Measures at Yale University and one owned by this writer are almost identical in size, form, and details. The only variation is that the Yale example (fig. 67) has a bubble level under a brass strip set into one end, an item lacking in the other example (fig. 68).

The compass card, made from a line engraving, is identical in each of the three examples. A floriated fleur-de-lis on the North

[125] Charles W. Brewster, *Rambles about Portsmouth* (Portsmouth, N. H.: L. W. Brewster, 1859, 1873), ser. 1, pp. 165, 329.

[126] Charles W. Brewster, *Rambles about Portsmouth* (Portsmouth, N. H.: L. W. Brewster, 1869), ser. 2, pp. 27, 90, 93, 136, 233, 263, 277, 316, 322, 367.

Figure 67.—Wooden surveying compass made by Thomas Salter Bowles of Portsmouth, New Hampshire. With spirit level. Made of birch, the compass is 13 in. long and has a diameter of 6 in. In the Streeter Collection of Weights and Measures, Yale University.

point has a compass and square at its base, and the name T. S. BOWLES is on a riband over it. Adorning the East point is an American eagle bearing a shield with stars and stripes and clutching arrows in one claw and a laurel twig in the other. In a ring within the central medallion is inscribed (see fig. 68), "* T. S. BOWLES * PORTSMOUTH, N.H. *"

The most interesting of the three instruments was acquired by the Dartmouth Museum as part of a collection of the late Frank C. Churchill, an inspector in the Indian Service. The instrument (fig. 69) is a quarter circle with a compass in its center and sighting bars mounted on a swinging arm that reads the angle of the brass scale on the arc by means of a vernier. It is mounted on a wooden tripod with the customary ball-and-socket joint, which permits it to be placed on a vertical plane. A built-in plumb bob at the side helps to establish the vertical.[127]

Interesting features of this instrument are two inscriptions engraved on the brass strip on the top of the dial. One states that it was "INVENTED BY P. MERRILL ESQ." and the other relates that it was "MADE BY JOHN KENNARD NEWMARKET." No information about P. Merrill has been found, and it is presumed that it was he who conceived the idea of combining the various elements into a single instrument and that it was made under his direction by Kennard.

[127] Information from Prof. Alfred F. Whiting, Dartmouth College Museum.

Figure 68.—Wooden surveying compass made by Thomas Salter Bowles (1765/70–post 1821) of Portsmouth, New Hampshire. Made of walnut, it is 12 in. long and has a diameter of 5⅜ in. With walnut sighting bars. In collection of writer.

Made By John
Newmarket

Figure 69.—Wooden surveying instrument inscribed "Invented by P. Merrill, Esq." and "Made by John Kennard, Newmarket." Made of walnut, 7¾ in. long; in its original pine case, with cover. The compass card and dial (see opposite) were made by Thomas Salter Bowles of Portsmouth. In Frank C. Churchill Collection, Dartmouth College Museum, Hanover, New Hampshire.

Some data on Kennard is available in a history of Newfields (formerly Newmarket) by Reverend Fitts. John Kennard was born in Kittery, Maine, in 1782. He learned the trade of clockmaker in Portsmouth, New Hampshire, presumably working with the members of the Ham family or others. On July 3, 1806, he married Sarah Ewer. He lived for various periods in Nashua and Concord before moving to Newfields in 1812. He lived in the Palmer house (which was burned in September 1899), and he kept a store in the little community and also served as its postmaster from 1822 to 1824. The post office was the only public office in the town until the cotton mills were built on the Lamprey River in 1823. Kennard later built and occupied the Kennard house on Piscassic Street, which was subsequently owned by Jeremiah Towle and has since been burned. In December 1830 he established an iron foundry together with Temple Paul and the Drake family, but in 1834 he sold his interest to Amos Paul and others. He was the father of six children and he died in 1861. During his lifetime he had specialized in making tall case and banjo clocks.[128]

[128] Rev. James Hill Fitts, *History of Newfields, New Hampshire, 1638–1911*, (Concord: Rumford Press, 1912).

The New Era

THE BEGINNING OF THE 19th century saw increased trading and shipping resulting from the economic development of the new republic, and the westward surge brought increased preoccupation with the settlement of communities and the development of land areas. As a consequence, the demand for instruments likewise increased.

Whereas during the 18th century and until some time after the end of the Revolutionary War probably not more than a dozen instrument makers and dealers are known to have emigrated from England or elsewhere to make their homes and careers in the American Colonies, the beginning of the 19th century saw substantial numbers of English and French instrument makers and dealers immigrate to the United States, to establish shops in the major centers of trade.

And whereas the names of scarcely a hundred mathematical-instrument makers who worked in the American Colonies during the 18th century are known today, the names of hundreds of similar 19th-century craftsmen and dealers are to be found.

As Derek Price [129] has so cogently stated: "For scientific instrument makers, one need only examine the nineteenth century city directories of Boston, Philadelphia and New York to find hundreds of names of craftsmen and firms. It is, to be sure, an antiquarian research, for one does not expect to find great discoveries coming from these people. But just as in Europe, it is a populous trade, influential in the growth of science and highly effective in spreading and intensifying the itch for ingenious instruments and devices. It is by these men that the basic skills of the Industrial Revolution were populated. . . ." By such means did American science and technology come of age.

---

[129] PRICE, op. cit. (footnote 1), p. 64.

The National Collection

Early American Scientific Instruments and Related Materials in the United States National Museum, Listed by Makers and Users

ADAMS, GEORGE; Fleet Street, London.
(See Ellicott, Andrew; Surveying Instrument.)

BARDIN, W. & T. M.; 16 Salisbury Square, Fleet Street, London.
(See Priestley, Joseph: Globes.)

BENNET, N. (fl. 1777); Middleboro, Mass., or Middleboro, Pa.

Alidade, plane table, scale 7⅞ in. radius, compass 5⅜ in. long. Brass scale and sights with compass in wooden box. Instrument inscribed "N. Bennet—Middlebor 1777." Although the name of this instrument maker does not appear on list of English or American makers, it is believed that he was American.

USNM 319076.

ELLICOTT, ANDREW (1754–1820); Baltimore, Md.

Instrument Box for astronomical instruments. Made of rosewood, with a hinged top, green felt underlining, brass lock, size 3 in. by 3 in. by 11 in. Owned and used by Andrew Ellicott for storage and transportation of small astronomical equipment.

Gift of John E. Reynolds, Ellicott's great-grandson, of Meadville, Pa., in 1932. USNM 310418.

Journal and *Astronomical Notebook*, manuscript written by Andrew Ellicott while locating the U.S. boundary line between the United States and the Spanish territory of Florida, 1797–1801. Contains day-by-day entries of experiences, field notes, and calculations made by Ellicott. The major part of the manuscript was published in *The Journal of Andrew Ellicott.*[130] Bound volume with

[130] The full title is *The Journal of Andrew Ellicott, Late Commissioner on behalf of the United States During Part of the Year 1796, the Years 1797, 1798, 1799 and Part of the Year 1800 For Determining the Boundary Between the United States and the Possessions of His Catholic Majesty in America.* It was published by Budd and Barton for Thomas Dobson at "the Stone House, No. 41 South Second Street" in Philadelphia in 1803.

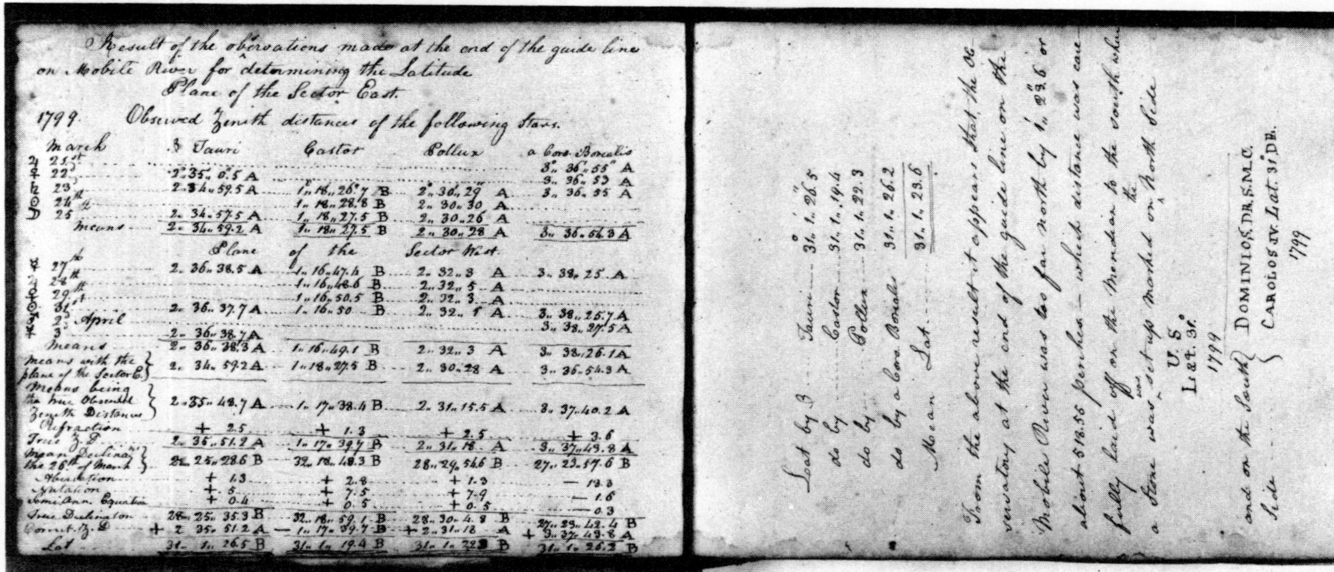

Figure 70.—Pages from manuscript "Journal and Astronomical Notebook" (USNM 310417) written by Andrew Ellicott while locating the boundary between the United States and the Spanish territory of Florida. These pages relate to the observations made in 1799 at the cord of the guide line on Mobile River for determining the latitude.

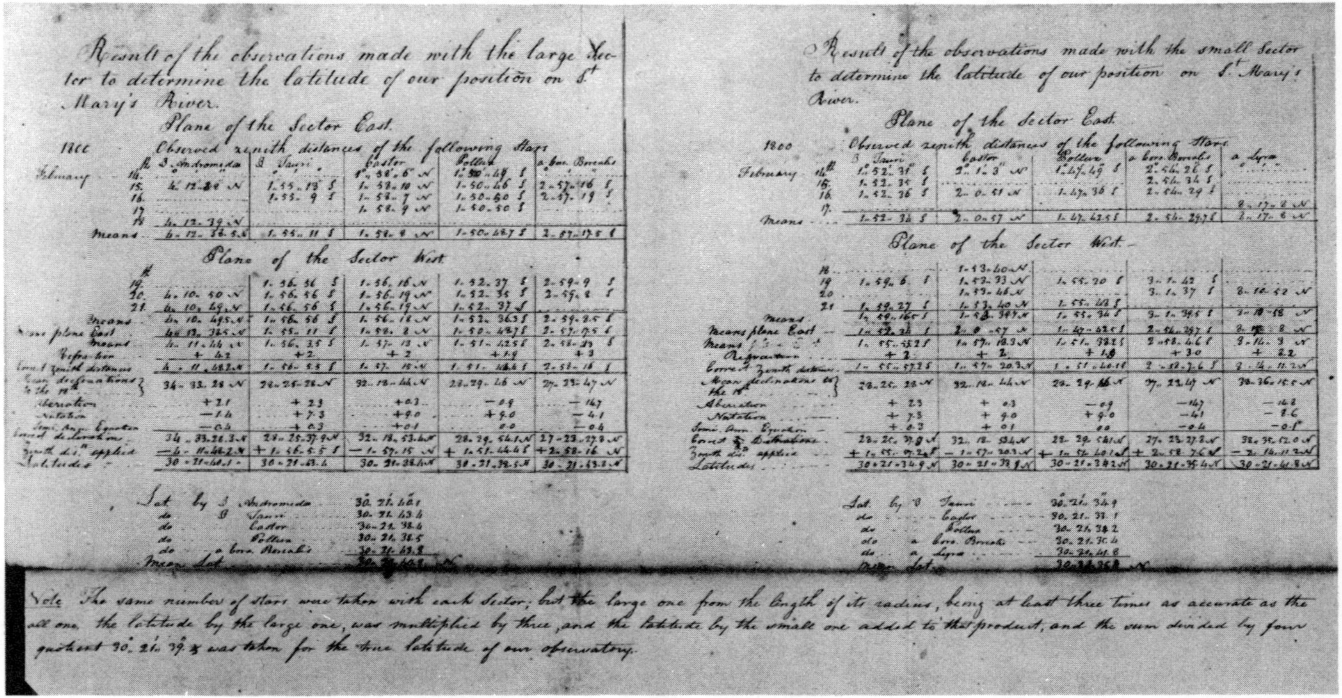

Figure 71.—Folding plate from Andrew Ellicott's "Journal and Astronomical Notebook" (USNM 310417), relating the results of observations made in February 1800 with the large and small sectors for determining Ellicott's position on St. Mary's River.

brown leather covers, end opening, marked "And. Ellicott," 6½ in. by 8 in. by 2 in. First page has signature "Andrew Ellicott 1788."

Formerly the property of Ellicott's eldest daughter, Jane Judith Ellicott, from whom it passed to her youngest son, William Reynolds. It was inherited by the latter's son, John Reynolds of Meadville, Pa., who presented it as a gift to the U.S. National Museum in 1932. USNM 310417. FIGURES 70, 71.

Pocket Slate 7¼ in. long and 4 in. wide, with wooden frame 7¼ in. long and 4 in. wide. Slate 5¾ in. long and 2½ in. wide. Part of field equipment used by Ellicott.

Gift of Charles Ellicott of Dansville, N.Y., in 1960. USNM 318292.

Quadrant of brass made and used by Ellicott. Quadrant has a radius of 12 in., is on a stand 17 in. high, and has the original lenses. Simple construction with easy adjustment, accomplished by means of two plumb lines. A tangent screw for slow motions was designed and added in 1885 by Andrew Ellicott Douglass, Ellicott's grandson. Instrument was made by Ellicott about 1790 and was used in running the southern boundary of the United States in 1796 and 1800, and on other surveys.

Deposit of Andrew Ellicott Douglass of Tucson, Ariz., in 1931. USNM 152081. FIGURE 72.

Surveying Instrument, with brass disk 10½ in. in diameter laid off in degrees, minutes, and seconds with vernier points. Two telescopes, one fixed and the other revolving. The instrument is mounted on a tripod or Jacob's staff by means of a socket on the underside. Complete with original painted pine case. The name of the maker, "G. Adams London," is engraved on the dial.

George Adams (1704–1773) was mathematical instrument maker to King George III. After serving an apprenticeship from 1718, he made instruments for the East India Company in 1735 and 1736, and established a shop at "Tycho Brahe's Head" at the corner of Raquet Court, Fleet Street. He specialized in terrestrial and celestial globes and microscopes. Following his death he was succeeded in business by his son George Adams the Younger (1750–1795), who also served as mathematical-instrument maker to the king.

This instrument is believed by the donor to have been used by either Andrew Ellicott or by his son-in-law David Bates Douglass.

Gift of Charles B. Curtis of Litchfield, Conn., in 1945. USNM 312932.

Figure 72.—Brass quadrant made by Andrew Ellicott about 1790 and used for running the southern boundary of the United States about 1796 and 1800, and on later surveys. USNM 152081.

Telescope, consisting of a brass tube 3½ in. long with an aperture of 2¾ in.; on its original brass tripod, with a serviceable altazimuth mounting. Late 18th century. Made by "W. & S. Jones/135 Holborn/London."

The firm of "W. & S. Jones" was a partnership of two brothers, Samuel and William Jones, opticians, who worked at 30 Lower Holborn and at 135 Holborn in London, from 1793. They bought the copyright to the books of George Adams, and subsequently largely carried on the original business of the Adams instrument makers.

In *The Journal of Andrew Ellicott* its author describes this instrument as the first of "Two Acromatic Telescopes for Taking signals, with sliding tubes, one of them drew out to upwards of 4 feet, and the other to about 15 inches, the latter for its length is remarkably good, it shows the satellites of Jupiter very distinctly."

Deposit of Andrew Ellicott Douglass of Tucson, Ariz., in 1899. USNM 152082. FIGURE 73.

Telescope, draw type, made of brass with acromatic lens, length 11 in. Incomplete, and maker not known. The second of the instruments described in *The Journal of Andrew Ellicott* as an acromatic telescope. Used for taking signals, with sliding tubes, which draw out to about 15 in. It was considered to be remarkably good for its length, and showed the satellites of Jupiter very distinctly.

Gift of Andrew Ellicott Douglass of Flagstaff, Ariz., in 1931. USNM 152085.

Transit and Equal Altitude Instrument, made entirely of brass, with original lens now broken. The instrument is described by Ellicott in the following extract from *The Journal of Andrew Ellicott:*

> Preparatory to beginning the ten mile square [of Washington] a Meridian was traced at Jones' Point on the West of the Potomac. From this Meridian an angle of 45 degrees was laid off North Westerly and a straight line continued in that direction ten miles. . . . From the termination of this second line a third making a right angle with it was carried South-Easterly ten miles: and from the beginning on Jones' Point a fourth was carried ten miles to the termination of the third. These lines were measured with a chain which was examined and corrected daily, and plumbed whenever the ground was uneven, and traced with a transit and equal altitude instrument which I constructed and executed in 1789 and used in running the Western boundary of the State of New York. This instrument was similar to that described by Le Monnier in his preface to the French "Histoire Celeste." . . . All the lines in this city in which I have been concerned were traced with the same instrument which I used on the lines of the ten mile square but as the Northern part was not finished when I left that place, I cannot pretend to say what method has since been pursued.

Deposit of Andrew Ellicott Douglass of Flagstaff, Ariz., in 1931. USNM 152080. FIGURE 10.

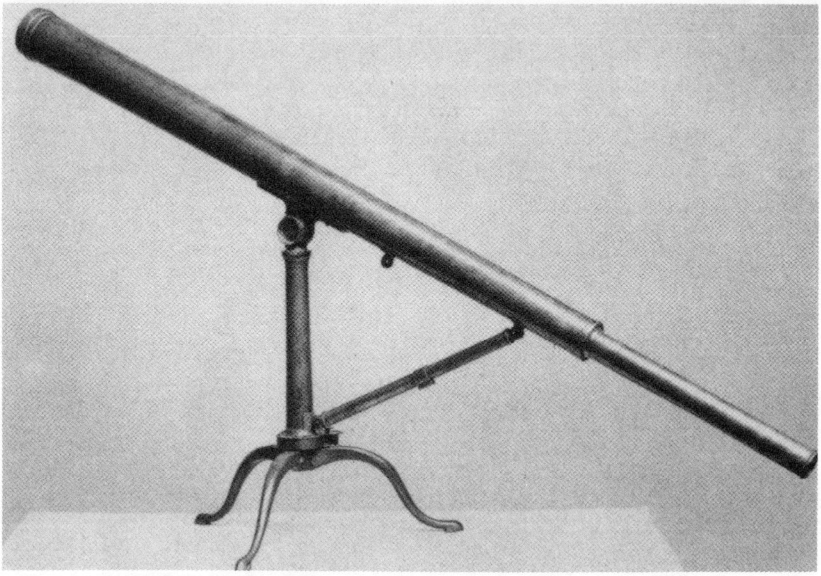

Figure 73.—Telescope used by Andrew Ellicott for his survey of the boundary between the United States and the Spanish territory of Florida. The instrument is signed "W. & S. Jones, 135 Holborn, London." USNM 152082.

ELLIS, ORANGE WARNER (18th century).

Theodolite, about 1780, brass; horizontal circle 5 in., vertical circle 5 in., telescope 7½ in., compass 3 in.; spirit level set into compass card; spirit level attached to telescope; fixed vertical circle; unsigned. Used by Orange Warner Ellis about 1780 in the surveying of the boundary between the United States and Canada, the area which is now Vermont.

Acquired from Miss Mary N. Ellis of Chicago, Ill., in 1929. USNM 309596. FIGURE 74.

FRYE, JOSEPH (fl. 1762–1783), Fryeburg, Maine.

Manuscript Booklet of "Tables Useful in Surveying Land, made and presented by Joseph Frye to his son, Joseph Frye, Jr., November 18, A. D. 1783." Size 6¼ in. by 3⅞ in., 16 pages, paper covers, marked "Fryeburg Joseph Frye AD MDCCLXXXIII."

Loan from Laurits C. Eichner of Clifton, New Jersey, in 1957. USNM 315062. FIGURE 45.

(See Greenough, Thomas, for surveying compass used by Joseph Frye.)

Figure 74.—Theodolite used by Orange Warner Ellis about 1780 for surveying boundary between the United States and Canada in the area which is now Vermont. USNM 309596.

GREENOUGH, THOMAS (1710–1785), Boston, Mass.

Surveying Compass, made of hickory with engraved paper compass card. Over-all length 11 ft.; dial 5½ in. in diameter. Central medallion on card depicts man along shoreline using a Davis quadrant with a schooner offshore, with touches of red. Inscribed in gilt in band around central medallion: "Made and Sold by THOMAS GREENOUGH, Boston, New Eng." Used by Joseph Frye in 1762 for surveying his land grant in what is now Fryeburg, Maine. Loan from Laurits C. Eichner, Clifton, N.J., in 1957. USNM 315001. FIGURE 44.

(See also, Frye, Joseph, manuscript booklet of "Tables Useful for Surveying Land . . .")

HAGGER, WILLIAM GUYSE (C. 1748?–1830?), Newport, R.I.
Backstaff, or *Davis Quadrant*, about 1760–1770, made of dark wood with scales and sights of boxwood, 25 in. long, 14 in. wide at large arc and 5 in. wide at small arc. Inscribed as follows: "Wm G. Hagger Newp^{t} R. Island/ For M^{r} ——." The name of the original owner has been blocked out by the insertion of a piece of ivory. This quadrant was acquired from Mrs. Carola Paine of Bethel, Conn., in 1961. USNM 319029. FIGURE 59.

Davis quadrants signed by Hagger are in the Comstock Memorial Collection of the Rhode Island Historical Society (dated 1776); in the Shepley Library in Providence, R.I. (dated 1768); and in the Peabody Museum at Salem, Mass. (dated 1775).

Also in the U.S. National Museum is an unsigned quadrant (USNM 178975) that is almost identical in detail to the one signed by Hagger. It is the gift of A. R. Crittenden, Middletown, Conn. Another almost identical instrument, in the collection of the Franklin Institute, is signed "C. Elliott, New London, 1764"; it differs from the other two only in that a lens is combined in the middle sight.

HOLBECHER, JOHN, (fl. 1738).
Backstaff, or *Davis Quadrant*, of dark wood with boxwood scales and vanes. Length 25½ in.; large arc 15 in. Inscribed "Made by John Holbecher/ For Capt. Joseph Swan—1738."

Holbecher is not listed as an English or American instrument maker, but it is believed that the instrument is American.

Acquired from Bern C. Ritchie & Co., Chicago, Ill., in 1960. USNM 318439.

JOHNSON, JOHN, Surveyor, 1818.
(See Rittenhouse & Evans, surveying compass.)

JONES, W. & S., 135 Holborn, London.
(See Ellicott, Andrew, telescope.)

PIERCE, ABNER (c. 1790).
Surveying Compass with Jacob's staff. Made of brass; 12 in. long; 5 in. in diameter; with needle lift. Jacob's staff 4 ft. high and with wood shaft about 1½ in.; brass head. Unsigned. Used about 1790 by Abner Pierce, who built Pierce's Mill in Rock Creek, District of Columbia.

Gift of Mrs. Francis D. Shoemaker of Washington, D.C., in 1930. USNM 309826.

Priestley, Joseph (1733–1804), Northumberland, Pa.

Chemical Apparatus that formed part of the laboratory of Joseph Priestley at his home. It includes the following specimens: 3 chemical retorts, 6 bell jars, 1 gas collecting flask, 6 flasks, 4 funnels, 23 miscellaneous metal and glass objects, and 1 eudiometer. A special exhibition of some of this chemical apparatus was held in the U.S. National Museum in 1958 (see fig. 69).

Gift of Miss Frances D. Priestley of Northumberland, Pa., in 1958. USNM 315341–315358. Figure 75.

Globes, one terrestrial (fig. 76) and one celestial (fig. 77), that formed part of the equipment used by Dr. Joseph Priestley. The terrestrial globe, of 26 in. diameter, has a Sheraton mahogany tripod stand and is inscribed—

> To the Rt. Honorable/Sir Joseph Banks, Bart. K.B./President of the Royal Society/containing all the latest Discoveries and Communications from the most/correct surveys to the year 1798/by Capt. Cook and more recent Navigators. Engraved upon an accurate degree by Mr. Arrowsmith,

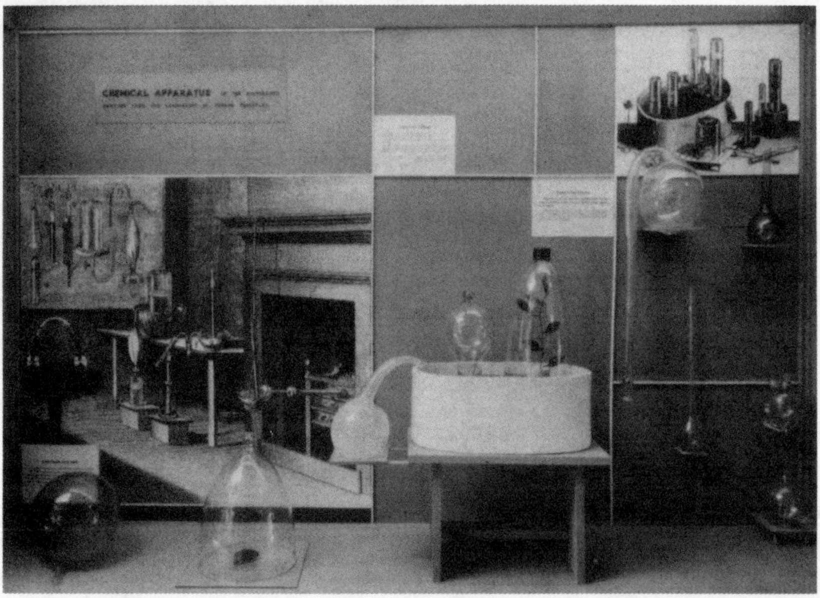

Figure 75.—Special exhibition of chemical laboratory apparatus used by Dr. Joseph Priestley. USNM 315341–351358.

Figure 76.—Terrestrial globe made by W. & T. M. Bardin of London and used by Dr. Joseph Priestley. Diameter, 26 in. USNM 53253.

Geographer/Respectfully Dedicated/by his most obedient servant/W. & T. M. Bardin/Manufactured and Sold Wholesale and Retail by W. & T. M. Bardin/16 Salisbury Square/Fleet Street, London.

The celestial globe, also with a Sheraton mahogany tripod stand, has a diameter of 23 in. and is inscribed—

To the Rev./Nevil Maskelyne, D. D. F. R. S./Astronomer Royal/This New British Celestial Globe/containing the positions of nearly 6,000 stars, clusters, nebulae, Planetary Nebulae/& correctly computed & laid down for the year 1800 from the latest observations and discoveries by Dr. Maskelyne, Dr. Herschel, the Rev. Mr. Wollaston, etc., etc./Is respectfully dedicated by his most obedient hmbl Servants W. & T. M. Bardin, Manufactured and sold Wholesale & Retail by W. & T. M. Bardin/16 Salisbury Square/Fleet Street, London.

Gifts of Mrs. Eliza R. Lyon of Williamsport, Pa., in 1893. USNM 53253, 53254. FIGURES 76, 77.

Orrery, mounted on three legs 31 in. high, round top 22½ in. in diameter. The planets shown are Mercury, Venus, Mars, Earth, Jupiter, and Saturn. The base is not original. Maker not known; English, 18th century.

Gift of Miss Frances Priestley of Northumberland, Pa., in 1958. USNM 315353. FIGURES 76, 77.

RITTENHOUSE, BENJAMIN (1740–c.1820).

Surveying Compass, about 1796, of brass, 13½ in. long over-all and 6½ in. diameter. Supported on a tripod by means of a ball-and-socket joint and screw-tightening device. The name "A. Ellicott" is inscribed on one arm outside the bezel of the dial, and the name "B. Rittenhouse" is inscribed on the other arm. The number "10" is marked on the reverse of this instrument, which is listed in the *Journal of Andrew Ellicott* as Item 9: "A Surveying Compass made by Mr. Benjamin Rittenhouse upon the newest and most approved plans."

Gift of Henry B. Douglass of Newton, N.J., in 1934. USNM 310815. FIGURE 78.

RITTENHOUSE, DAVID (1732–1796), Philadelphia, Pa.

Surveying Compass, brass, over-all length 14 in., diameter 6½ in., silvered dial marked with eight-pointed star indicating the cardinal and intermediate points, glazed. Inscribed "Rittenhouse, Philadelphia." Fitted with a ball-and-socket joint for mounting on a tripod, and complete with wooden field case.

Stated to have been used by General Washington for laying out

Figure 77.—Celestial globe made by W. & T. M. Bardin of London and used by Dr. Joseph Priestley. Diameter, 23 in. USNM 53254.

Figure 78.—Brass surveying compass made by Benjamin Rittenhouse for Andrew Ellicott and inscribed with both names. The instrument is described in *Journal of Andrew Ellicott* (Philadelphia, 1803). USNM 310815.

the estate of Mount Vernon, according to family manuscripts. It was made by David Rittenhouse and presented by him to General Washington, who subsequently gave it to Capt. Samuel Duvall.

A manuscript consisting of 14 letters relating to the surveying compass is filed in the U.S. National Museum (USNM 92542). The letters were written in 1851 and 1852 by George Washington Parke Custis, Anthony Kimmel, and other Washington descendants.

Gift of Anthony Kimmel to the U.S. Government, and transferred to the U.S. National Museum in 1883. USNM 92538. FIGURE 79.

Zenith Sector for measuring the angle between a star at its zenith and the vertical. Made of brass, with focal length of 6 ft. and an aperture of 2½ in. The original lens was made in London about 1780. The instrument was made in the old pattern with brass tube and mountings and a wooden supporting post. The tube is suspended by trunnions at the top and swings against a graduated arc extending north and south for measuring zenith distances in the meridian. It is adjusted in the vertical by a plumb line whose errors are eliminated by reversing the whole mounting about the supporting post. Constructed principally by David Rittenhouse, with some modifications by Andrew Ellicott.

Figure 79.—Surveying compass made by David Rittenhouse for Gen. George Washington, inscribed "Rittenhouse, Philadelphia." This instrument was used by Washington in making a complete survey of his estate at Mount Vernon, 1796–1799. The survey was assisted by Capt. Samuel Duval, surveyor of Frederick County, Maryland. Washington gave the instrument to Captain Duval, from whom it descended to the Hon. Anthony Kimmel, who donated it to the U.S. National Museum. USNM 92538.

In the *Journal of Andrew Ellicott* its author referred to this sector as follows:

> The boundary line to the North of Pennsylvania was fixed by Dr. Rittenhouse and Captain Holland in the year 1774 and completed in 1786 and 1787. We commenced operations by running a guide line west from the point mentioned on the Delaware 20¼ miles and there corrected by the following Zenith distances taken at its West termination by a most excellent sector constructed and executed by Dr. Rittenhouse.

The zenith sector is again mentioned in the appendix of the *Journal:* "One Zenith Sector of nearly six feet radius similar to the one made by Mr. [George] Graham for Dr. Bradley and Mr. Molyneux, with which the aberrations of the stars and mutation of the earth's axis were discovered, and the quantities determined."

Gift of Andrew Ellicott Douglass, Tucson, Ariz., in 1931. USNM 152078. FIGURE 11.

Zenith Sector, made of brass, original lens broken. Constructed by David Rittenhouse with some additions made by Andrew Ellicott. In *The Journal of Andrew Ellicott* the instrument is described as a—

. . . Zenith Sector of 19 inches radius to be used when the utmost accuracy was not necessary, and where the transportation of the large one could not be effected without great expense and difficulty. These instruments were principally executed by my late worth and ingenious friend, Mr. Rittenhouse, except some additions which I have made myself. The plumb lines of both Sectors are suspended from a notch above the axis of the instruments in the manner described by the Rev. Dr. Maskelyne, the present Astronomer Royal at Greenwich, in the introduction to the first volume of his Astronomical Observations. A particular description of those instruments is rendered unnecessary by being accurately done in a number of scientific works, particularly by M. de Maupertius in his account of the measurement of a degree of the meridian under the Arctic Circle—The Sector is of all instruments the best calculated for measuring zenith distances which come within its arc. The large one above mentioned [large Zenith Sector] extends to 5 degrees North, and South of the Zenith. Stars when so near the Zenith are insensibly affected by the different refractive powers of the Atmosphere arising from its different degrees of density. Add to this that the error of the visual axis is completely corrected by taking the Zenith distances of the stars with the plane, or face of the instrument both East and West.

USNM 152079. FIGURE 80.

RITTENHOUSE & EVANS, Philadelphia, Pa., 18th century.

Surveying Compass, about 1780, made of brass, overall length 13¾ in., diameter of dial 5¼ in., silvered bubble level, vernier on alidade. The glazed dial, engraved "Rittenhouse & Evans," is fitted with a brass cover.

This instrument was made during a brief partnership between David Rittenhouse and David Evans, a clock- and watchmaker of Philadelphia and Baltimore. It was one of several owned and used by John Johnson in 1818 for surveying the boundaries between Canada and Maine.

The survey, made in compliance with the Treaty of Ghent, is described in *The Collections of the Maine Historical Society* (Portland: Hoyt, Fogg & Donham, 1881, vol. 8, p. 20):

Thomas Barclay, of whom we have heard more than once before, as a Commissioner under the treaty, on the part of Great Britain, and Cornelius P. Van Ness, on the part of the United States, were appointed Commissioners to ascertain and run the line. An actual survey was arranged, and surveyors appointed, to wit: Charles Turner, Jr., on the part of the United States, and Colin Campbell on the part of Great Britain. About twenty miles of the line was surveyed, then the work was discontinued, never to be resumed; but an exploring survey was commenced by Colonel Bouchette, on the part of Great Britain, and John Johnson, on the part of the United States. These gentlemen made an exploring line in 1817, extending ninety-nine miles from the monument at the head of the river St. Croix, and made separate reports of their doings. In 1818 Mr. Johnson, with Mr. Odell, who had taken the place

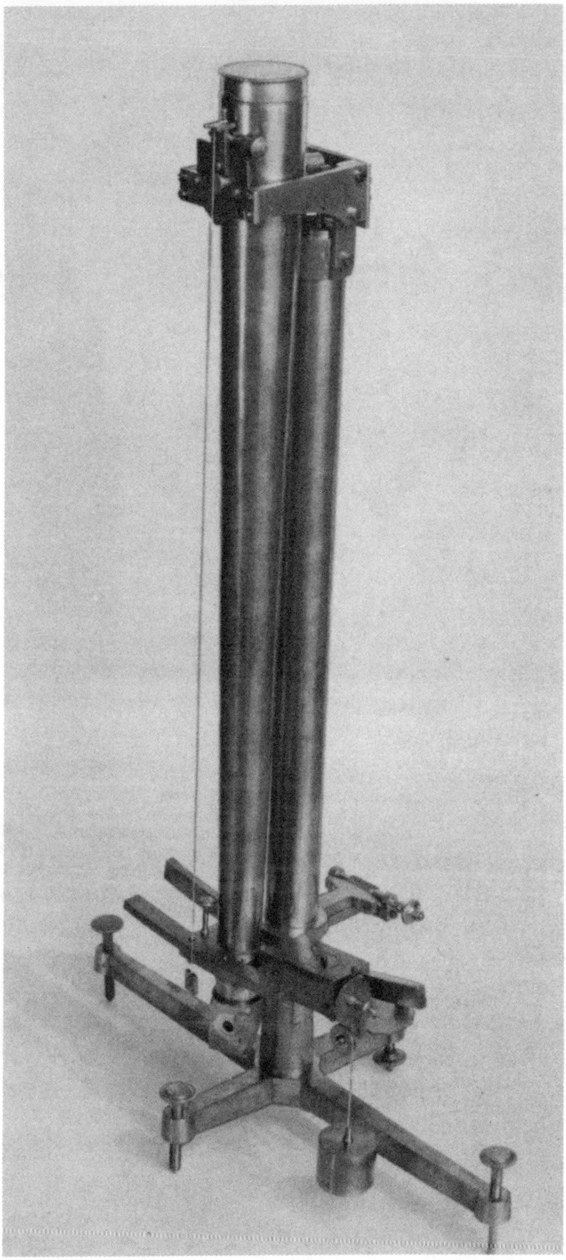

Figure 80. Zenith sector, with a radius of 19 in., constructed by David Rittenhouse for Andrew Ellicott. USNM 152079.

Figure 81.—Brass surveying compass marked "Rittenhouse & Evans," about 1780. Over-all length, 13¾ in.; diameter of dial, 5¼ in. This instrument, made about 1780, was owned and used by John Johnson in 1818 for surveying the boundaries between Canada and Maine. USNM 309543.

of Col. Bouchette, finished running the exploring line to the Beaver or Metis River. . . .

Gift of John Johnson Allen of Burlington, Vt., in 1927. USNM 309543. FIGURE 81.

THOMPSON, Captain SAMUEL ROWLAND (18th century); Lewes, Del.
Octant made of dark wood and with lignum vitae; brass fittings. This harbormaster's instrument, used by Captain Thompson during the second half of the 18th century, is without numerical designations on the arc. The eighth part of a circle is connected to an apex by two side pieces with a swinging arm hinged at the apex, with a blade at its end that moves along a checkered scale on the arc.

Gift of George Andrews Thompson of Baltimore, Md., in 1926. USNM 308473.

VOIGHT, HENRY (1738–1814), Philadelphia, Pa.
Equal Altitude Telescope of brass, 17 in. long, on wooden tripod about 46 in. high. Objective lens is missing. Signed "Henry Voigt." Made about 1790 and used for determining meridian lines and time observation of the sun's noon transit. This form of

instrument was originally invented about 1716 by Roger Cotes, professor of astronomy at Cambridge, as a simple instrument for the determination of time.

Deposited in the U.S. National Museum by the Smithsonian Institution in 1939. USNM 311772. FIGURE 31.

WASHINGTON, General GEORGE (1732–1799), Mount Vernon, Va. *Compass Sundial* described by the donor as having been presented to Gen. George Washington by General Braddock on the retreat through Paris Gap, Fairfax County, Va. Gift of Samuel Keese in 1902. USNM 9842.

Field Glass, brass tube in three sections, length closed 9 in., opened 22½ in. Diameter of object lens 1¾ in., of ocular lens 1⅛ in. With original case of russet leather, which is 9½ in. long and 2½ in. in diameter. Maker not known. Stated to have been used by Washington during the Revolutionary War at the campaign of Valley Forge.

According to related correspondence, when not in use the instrument was carried by the General's body servant, Billy Lee. The General presented the field glass to Major Lawrence Lewis, his favorite nephew, in 1799, the last year of his life.

Purchased by the U.S. Government from the Lewis heirs in 1878 and transferred to the U.S. National Museum in 1883. USNM 92424, 92425. FIGURE 82.

Spyglass or *Telescope*, made of wood, 9-sided, wrapped throughout with twine, 62 in. long. Brass mountings for object and ocular lenses made by "Cole, Fleet Street, London." Diameter of object lens 2¾ in., diameter of ocular lens 1 in.

The maker, Benjamin Cole (1725–1813), was the third generation of instrument makers of the same name. Other instruments

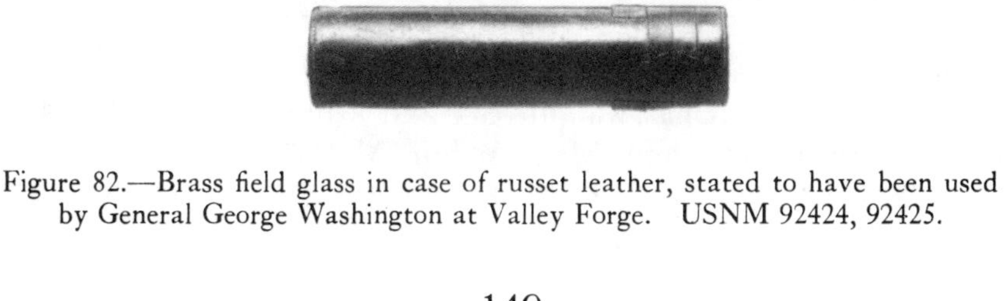

Figure 82.—Brass field glass in case of russet leather, stated to have been used by General George Washington at Valley Forge. USNM 92424, 92425.

Figure 83.—Telescope, 62 in. long, made of wood wrapped with twine. It was made by Benjamin Cole of London and was owned and used by Gen. George Washington at Mount Vernon. USNM 92423.

by this maker are in the National Maritime Museum and the Whipple Museum, Cambridge.

This telescope, used by General Washington at Mount Vernon, "was kept behind the hall door and his favorite amusement was to look out over the river with it." According to Mrs. Lewis, the General used it to observe life on the river and especially to discover guests approaching Mount Vernon, as many of their visitors arrived by boat. Benjamin Latrobe, the architect, on a visit to Mount Vernon made an amusing sketch of his host looking anxiously up the stream for some belated dinner guests.

Part of the collection purchased from the Lewis heirs in 1878 by the U.S. Government and transferred to the U.S. National Museum in 1883. USNM 92423. FIGURE 83.

Survey of Land, drawn and documented by George Washington on April 2, 1751 for Thomas Loftan of Frederick County, Va. Paper, 12 in. wide by 7¾ in. high.

This survey was made by Washington when he was 19 years of age, and it is believed to be the only such document relating to his earliest period as a surveyor. Washington was licensed as a surveyor by the President and Masters of William and Mary College in 1749. On July 20th of the same year he was appointed surveyor in Culpepper County, Va., by Governor Dinwiddie.

Acquired in 1961. USNM 238367. FIGURE 84.

WHITE, PEREGRINE (1747–1834), Woodstock, Conn.

Surveying Compass, about 1790, made of brass, complete with original case, tripod, and gunter's chain. The instrument measures 12¼ in. overall. The dial, with a diameter of 5⅝ in. and a pewter vernier ring, is inscribed "PEREGRINE WHITE/Woodstock." Tripod is 57½ in. long and has walnut legs and a brass universal socket joint. Gift of Dr. and Mrs. Arthur M. Greenwood.

USNM 388993. FIGURE 23.

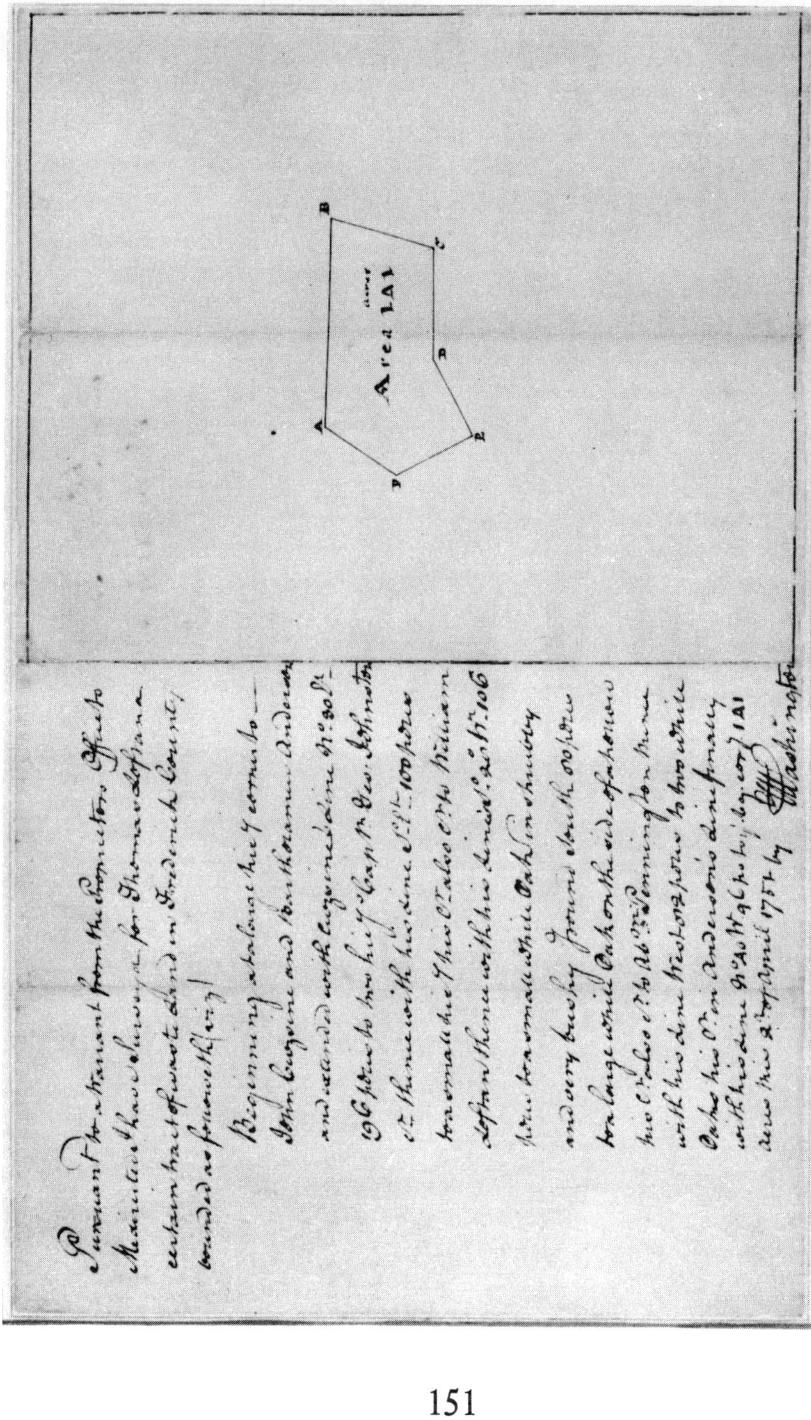

Figure 84.—Survey of land drawn and documented by George Washington for Thomas Loftan of Frederick County, Va., in 1751. Size: 12 in. wide, 7¾ in. high. USNM 238367.

WHITNEY, THOMAS (fl. 1798–1821), Philadelphia, Pa.
Pocket Compass of brass encased in brassbound mahogany box with separate carrying case. Paper dial is inscribed "T. Whitney/ Phil^{a}." Carried by Capt. William Clark on the Lewis and Clark Expedition to the Pacific Coast in 1803–1806.
USNM 38366. FIGURE 85.

Figure 85.—Pocket compass made and signed by Thomas Whitney of Philadelphia. With original carrying case. Carried by Capt. William Clark on the Lewis and Clark expedition to the Pacific Coast, 1803–1806. USNM 38366.

Appendix

SURVIVING WOODEN SURVEYING COMPASSES

(Asterisk denotes information unavailable)

Collection	*Type of wood*	*Length (in.)*	*Width (in.)*	*Height of bars (in.)*	*Length of needle (in.)*	*Maker and period*
Preston R. Bassett	Maple	9	5	3¼	*	Unsigned (18th century)
Bucks County Historical Society	Cherry	11	5½	6⅝	2⅜	Thomas Greenough of Boston (1710–1785)
Bostonian Society	Apple or walnut	13⅞	*	*	4¾	John Dupee of Boston (after 1761)
Dartmouth College Museum	Walnut	7¾	*	*	*	Thomas S. Bowles of Portsmouth, N.H. (c. 1765–1821)
	*	12	8	*	*	Unsigned (18th century)
	*	8⅜	4⅝	*	*	Unsigned (18th century)
L. C. Eichner (U.S. National Museum)	Hickory	11	5½	3	4	Thomas Greenough of Boston (1710–1785)
Farmer's Museum	Oak	12¾	6½	5	*	Unsigned (18th century)
Franklin Institute	Gum	13¾	5¾	4	5	Thomas Greenough of Boston (1710–1785)
Mariner's Museum	*	*	*	*	*	Charles Thacher (18th century)
Old Sturbridge	Maple	13	4	*	*	Unsigned (18th century)
	Maple	11⅝	5⅞	*	*	Thomas Greenough of Boston (1710–1785)
	Walnut	18	8	*	*	Aaron Breed of Boston (1791–1861)
New Hampshire Historical Society	Maple	11	5¾	2½	4⅝	Joseph Halsy of Boston (fl. 1697–1762)

Collection	*Type of wood*	*Length (in.)*	*Width (in.)*	*Height of bars (in.)*	*Length of needle (in.)*	*Maker and period*
N. Parker	Walnut	13½	4⅞	5	*	John Dupee of Boston (after 1761)
Peabody Museum	*	11	*	*	3	James Halsy II of Boston (1695–1767)
Worth Shampeny	*	*	*	*	*	Jedidiah Baldwin of Hanover, N.H. (c. 1777–1829)
South Natick Historical Society	Apple or walnut	13–16	*	*	4⅞	John Dupee of Boston (after 1761)
Streeter Coll., Yale University	Birch	13	6	*	4	Thomas S. Bowles of Portsmouth, N.H. (c.1765–1821)
	Cherry	11⁵⁄₁₆	6	4	*	Jere Clough of Boston (18th century)
	Cherry	12	6	3½	*	Benjamin Warren of Plymouth, Mass. (fl. 1740–1790)
Roleigh L. Stubbs	Cherry	7½	3¾	3	*	Unsigned
Silvio A. Bedini	Walnut	12	5⅜	5	4	Thomas S. Bowles of Portsmouth, N.H. (c. 1765–1821)
	Pine	5¾	3½	2½	*	Unsigned (18th century)
	Mahogany	*	*	*	*	Unsigned (18th century)
	Basswood	12	5¾	2¾	4	Thomas Greenough of Boston (1710–1785)
	Birch	18	7½	7½	6	Samuel Thaxter of Boston (1769–1842)
	Mahogany	13	7¼	4¼	6	Samuel Thaxter of Boston (1769–1842)
	Yellow birch	8¼	4	*	4¼	Benjamin K. Hagger of Boston and Baltimore (c. 1769–1834)
	Cherry	14	5½	6⅜	4¾	Gurdon Huntington of Windham, Conn. and Walpole, N.H (1763–1804)
Yale Gallery of Fine Art	Mahogany	11½	5	*	*	Andrew Newell of Boston (1749–c. 1798)

MATHEMATICAL PRACTITIONERS AND INSTRUMENT MAKERS

Alphabetical List

(Asterisk denotes information unavailable.)

Name	*Period*	*Place*	*Types of instruments*
Bailey, John	fl. 1778	Fishkill, N. Y.	Surveying; surgical
Bailey, John, II	1752–1823	Hanover and Lynn, Mass.	Surveying
Baily, Joel (practitioner)	1732–1797	West Bradford, Pa.	
Baldwin, Jedidiah	c. 1777–1829	Salem, Boston, and Northampton, Mass.; Hanover, N. H.	Surveying
Banneker, Benjamin (practitioner)	c. 1734–1806	Baltimore	
Benson, John	fl. 1793–1797	*	Optical
Biddle, Owen (practitioner)	1737–1799	Philadelphia	
Biggs, Thomas	fl. 1792–1795	New York and Philadelphia	Surveying
Blakslee, Ziba	1768–1834	Newtown, Conn.	Surveying
Blundy, Charles	fl. 1753	Charleston, S. C.	Thermometric; watches
Bowles, Thomas S.	c. 1765–1821	Portsmouth, N. H.	Surveying
Breed, Aaron	1791–1861	Boston	Surveying
Brokaw, Isaac	fl. 1771	Philadelphia	*
Bulmain & Dennies	fl. 1799	New York	Nautical
Burges, Bartholomew	fl. 1789	Boston	Scientific
Burnap, Daniel	1759–1838	East Windsor and Coventry, Conn.	Surveying; clocks
Caritat, H.	fl. 1799	New York	Astronomical
Chandlee, Benjamin, Jr.	1723–1791	Nottingham, Md.	Surveying; clocks
Chandlee & Bros.	fl. 1790–1791	Nottingham, Md.	Clocks; surveying
Chandlee, Ellis	1755–1816	Nottingham, Md.	Surveying; clocks
Chandlee, Ellis & Bros.	fl. 1791–1797	Nottingham, Md.	Clocks; surveying
Chandlee, Goldsmith	c. 1751–1821	Winchester, Va.	Surveying; astronomical; clocks
Chandlee, Isaac	1760–1813	Nottingham, Md.	Surveying; clocks
Clark, Robert	fl. 1785	Charleston, S.C.	Nautical; surveying; optical
Clough, Jere	18th century	Boston	Surveying

Name	*Period*	*Place*	*Types of instruments*
Condy, Benjamin	fl. 1756–1798, d. 1798	Philadelphia	Mathematical; sand glasses
Crow, George	c. 1726–1772	Wilmington, Del.	Surveying; clocks
Dabney, John, Jr.	fl. 1739	Boston	Mathematical
Dakin, Jonathan	fl. 1745	Boston	Mathematical; balances
Davenport, William	1778–1829	Philadelphia	Mathematical; surveying
Dean, William	(?–1797)	Philadelphia	Surveying; nautical
Devacht, Joseph and Francois	fl. 1792	Gallipolis, Ohio	Watches; compasses; sundials
Donegan (or Denegan), John	fl. 1787	New York	Glass; philosophical
Donegany, John (see Donegan)			
Doolittle, Enos	1751–1806	Hartford, Conn.	Surveying; nautical; clocks
Doolittle, Isaac	1721–1800	New Haven, Conn.	Clocks; scientific
Doolittle, Isaac, Jr.	1759–1821	New Haven, Conn.	Surveying; clocks
Dupee, John	fl. after 1761	Boston	Surveying
Ellicott, Andrew (also practitioner)	1754–1820	Baltimore	Surveying; astronomical
Emery, Samuel	1787–1882	Salem, Mass.	Mathematical
Evans, George	fl. 1796; d. 1798	Philadelphia	Mathematical
Fairman, Gideon (See Hooker and Fairman)	1774–1827	Newburyport, Mass.	Mathematical
Fisher, Martin	fl. 1790	Philadelphia	Glass
Folger, Peter (practitioner?)	1617–1690	Nantucket	
Folger, Walter, Jr.	1765–1849	Nantucket	Astronomical; surveying
Ford, George	fl. late 18th century to 1842	Lancaster, Pa.	Surveying; nautical
Ford, George, II	fl. 1842	Lancaster, Pa.	Surveying; nautical
Fosbrook, W.	fl. 1786 or earlier	New York	Surgical; dental
Gatty, Joseph	fl. 1794	New York and Philadelphia	Glass; philosophical
Gilman, Benjamin C.	1763–1835	Exeter, N.H.	Mathematical; clocks
Gilmur, Bryan	fl. end of 18th century	Philadelphia	Mathematical; clocks
Godfrey, Thomas	1704–1749	Philadelphia	Improved reflecting backstaff
Gould, John	fl. 1794	Philadelphia	Nautical; surgical; optical
Grainger, Samuel (practitioner)	fl. 1719	Boston	

Name	*Period*	*Place*	*Types of instruments*
Greenleaf, Stephen	1704–1795	Boston	Mathematical
Greenough, Thomas	1710–1785	Boston	Mathematical; surveying; nautical; astronomical
Greenough, William	fl. 1785	Boston	Surveying
Greenwood, Isaac, Sr. (practitioner)	fl. 1726	Boston	Surveying
Greenwood, Isaac, Jr.	1730–1803	Boston	Mathematical
Grew, Theophilus (practitioner)	fl. 1753	Philadelphia	
Hagger, Benjamin King	c.1769–1834	Boston and Baltimore	Mathematical; surveying
Hagger, William Guyse	c.1744–1830?	Newport, R.I.	Nautical
Halsie, James, I (practitioner)	fl. 1674	Boston	
Halsy, James, II	1695–1767	Boston	Mathematical; surveying
Halsy, John	fl. 1700	Boston	Mathematical
Halsy, Joseph	fl. 1697–1762	Boston	Surveying; nautical
Ham, James	fl. 1754–1764	New York and Philadelphia	Mathematical
Ham, James, Jr.	fl. 1780	Philadelphia	Mathematical
Hamlin, William	1772–1869	Providence, R. I.	Mathematical; nautical; astronomical
Hanks, Benjamin	1755–1824	Mansfield and Litchfield, Conn.	Surveying
Hanks, Truman	fl. 1808	Mansfield and Litchfield, Conn.	Surveying
Harland, Thomas	1735–1807	Norwich, Conn.	Surveying; clocks
Heisely, Frederick A.	1759–1839	Frederick, Md.; Lancaster, Harrisburg, and Pittsburgh, Pa.	Mathematical; surveying; clocks
Heisely, George	1789–1880	Harrisburg, Pa.	Clocks; mathematical
Hinton, William	fl. 1772	New York	Mathematical
Hoff, George	1740–1816	Lancaster, Pa.	Clocks; surveying
Holcomb, Amasa (also practitioner)	1787–1875	Southwick, Mass.	Surveying; astronomical
Hooker & Fairman (William Hooker and Gideon Fairman)	before 1810	Newburyport, Mass.	Mathematical
Houghton, Rowland	c. 1678–1744	Boston	Surveying
Huntington, Gurdon	1763–1804	Windham, Conn., and Walpole, N.H.	Surveying and other; clocks

Name	*Period*	*Place*	*Types of instruments*
Jacks, James	fl. 1780's	Charleston, S. C.	Mathematical; surveying
Jayne, John	late 18th century	Salem, Mass.	Mathematical
Kennard, John	1782–1861	Newmarket, N. H.	Surveying; clocks
Ketterer, Alloysius	fl. 1789	Philadelphia	Glass
King & Hagger (Benjamin King and William Guyse Hagger)	1759 or 1760 until early 1760's	Newport, R.I.	Mathematical; nautical
King, Benjamin	1707–1786	Newport, R.I.	Mathematical; nautical
King, Benjamin	1740–1804	Salem, Mass.	Nautical
King, Daniel	1704–1790	Salem, Mass.	Mathematical
King, Samuel	1748–1819	Newport, R.I.	Mathematical
Lamb, A. & Son	1780's	New York	Mathematical
Lamb, Anthony	1703–1784	England; Virginia; Philadelphia; New York; Hunter's Key, N.Y.	Mathematical; surveying; nautical
Lamb, John	1735–1800	New York	Mathematical
Mendenhall, Thomas	fl. 1775	Lancaster, Pa.	Mathematical; clocks
Miller, Aaron	fl. 1748–1771	Elizabethtown, N.J.	Surveying; clocks; compasses
Morris, M.	fl. 1785	New York	Protractors
Newell, Andrew	1749–1798	Boston	Mathematical; compasses
Newell, Joseph	fl. 1800–1813	Boston	Surveying
Pease, Paul	fl. 1750	Probably Rhode Island	Quadrant
Platt, Augustus	1793–1886	Columbus, Ohio	Mathematical; surveying
Platt, Benjamin	1757–1833	Danbury, Litchfield, and New Milford, Conn.; Lanesboro, Mass.; Columbus, Ohio	Compasses; surveying; clocks
Pope, Joseph	1750–1826	Boston	Scientific; clocks
Potter, John	fl. 1746–1818	Brookfield, Mass.	Surveying
Potts, W. L.	late 18th century	Bucks County, Pa.	Surveying
Prince, John (practitioner)	1751–1836	Salem, Mass.	Scientific
Prince, Nathan (practitioner)	fl. 1743	Boston	

Name	*Period*	*Place*	*Types of instruments*
Pryor, Thomas	fl. 1778	Philadelphia	Mathematical
Revere, Paul	1735–1818	Boston	Gunnery
Rittenhouse, Benjamin	1740–c.1820	Philadelphia	Astronomical; surveying
Rittenhouse, David (practitioner)	1732–1796	Philadelphia and Norriton, Pa.	Astronomical; surveying
Rittenhouse & Evans	fl. 1770's	Philadelphia	Surveying
Sibley & Marble (Clark Sibley and Simeon Marble)	late 18th century	New Haven, Conn.	Mathematical; clocks; watches
Smith, Cordial	fl. 1775	Connecticut	Surveying
Sommer, widow Balthaser	fl. 1753	New York	Optical
Sower, Christopher	c. 1724–1740	Germantown and Philadelphia, Pa.	Mathematical; clocks
Stiles & Baldwin (Jedidiah Baldwin)	fl. 1791	Northampton, Mass.	Surveying
Stiles & Storrs (Nathan Storrs and Jedidiah Baldwin)	fl. 1792	Northampton, Mass.	Surveying
Taws, Charles	fl. 1795	Philadelphia	Mathematical
Thacher, Charles	18th century	Probably Boston	Surveying
Thaxter, Samuel	1769–1842	Boston	Nautical; mathematical; surveying
Voight, Henry	1738–1814	Philadelphia	Astronomical; clocks; watches
Wall, George, Jr.	fl. 1788	Bucks County, Pa.	Surveying
Walpole, Charles	fl. 1746	New York	Mathematical
Warren, Benjamin	fl. 1740–1790	Plymouth, Mass.	Surveying; nautical
White, Peregrine	1747–1834	Woodstock, Conn.	Surveying; clocks
Whitney, John	fl. 1801	Philadelphia	Mathematical; optical
Whitney, Thomas	fl. 1798–1823	Philadelphia	Mathematical; optical; surveying
Williams, William	1737 or 1738–1792	Boston	Mathematical; nautical
Willis, Arthur (practitioner)	fl. 1674	Possibly Massachusetts	
Wilson, James	1763–1855	Bradford, Vt.	Globes
Wistar, Richard	fl. 1752	Wistarburg, N.J.	Glass
Witt, Christopher (practitioner)	fl. 1710–1765	Germantown, Pa.	Mathematical; clocks
Wood, John	fl. 1790	Philadelphia	Compasses
Youle, James	1740–1786	New York	Surgical
Youle, John	fl. 1786	New York	Surgical

MATHEMATICAL PRACTITIONERS AND INSTRUMENT MAKERS

Geographical Listing

CONNECTICUT

Coventry:	Daniel Burnap (1759–1838); surveying instruments and clocks.
Danbury:	Benjamin Platt (1757–1833); compasses and clocks.
East Windsor:	Daniel Burnap (1759–1838); surveying instruments and clocks.
Hartford:	Enos Doolittle (1751–1806); surveying and navigational instruments, compasses, and clocks.
Litchfield:	Benjamin Hanks (1755–1824); surveying instruments and clocks.
	Truman Hanks (fl. 1808); surveying instruments.
	Benjamin Platt (1757–1833); compasses and clocks.
Mansfield:	Benjamin Hanks (1755–1824); surveying instruments.
	Truman Hanks (fl. 1808); surveying instruments.
New Haven:	Isaac Doolittle (1721–1800); clocks and scientific instruments.
	Isaac Doolittle, Jr. (1759–1821); surveying instruments and clocks.
	Sibley & Marble (late 18th century); clocks and mathematical instruments.
New Milford:	Benjamin Platt (1757–1833); compasses and clocks.
Newtown:	Ziba Blakeslee (1768–1834); surveying instruments.
Norwich:	Thomas Harland (1735–1807); surveying instruments and clocks.
Windham:	Gurdon Huntington (1763–1804); clocks and surveying and other instruments.
Woodstock:	Peregrine White (1747–1834); surveying instruments and clocks.
———:	Smith, Cordial (fl. 1775); surveying instruments.

DELAWARE

Wilmington:	George Crow (c. 1726–1772); surveying instruments and clocks.

MARYLAND

Baltimore:	Benjamin Banneker (c.1734–1806), practitioner.
	Andrew Ellicott (1754–1820), practitioner; surveying and astronomical instruments.
	Benjamin K. Hagger (c.1769–1834); mathematical and surveying instruments.

Frederick:	Frederick A. Heisely (1759–1839); clocks and mathematical instruments.
Nottingham:	Benjamin Chandlee, Jr. (1723–1791); clocks and surveying instruments.
	Chandlee & Bros. (fl. 1790–1791); clocks and surveying instruments.
	Ellis Chandlee (1755–1816); surveying instruments and clocks.
	Ellis Chandlee & Bros. (fl. 1791–1797); clocks and surveying instruments.
	Isaac Chandlee (1760–1813); surveying instruments and clocks.

MASSACHUSETTS

Boston:	Jedidiah Baldwin (c.1777–1829); surveying instruments.
	Aaron Breed (1791–1861); surveying instruments.
	Bartholomew Burges (fl. 1789); scientific instruments.
	Jere Clough (18th century); surveying instruments.
	John Dabney, Jr. (fl. 1739); mathematical instruments.
	Jonathan Dakin (fl. 1745); mathematical instruments and balances.
	John Dupee (fl. after 1761); surveying instruments.
	Samuel Grainger (fl. 1719), practitioner.
	Stephen Greenleaf (1704–1795); mathematical instruments.
	Thomas Greenough (1710–1785); mathematical, surveying, astronomical, and nautical instruments.
	William Greenough (fl. 1785); surveying instruments.
	Isaac Greenwood, Sr. (c.1725–1750), practitioner.
	Isaac Greenwood, Jr. (1730–1803); mathematical instruments.
	Benjamin K. Hagger (c.1769–1834); mathematical and surveying instruments.
	James Halsie I (fl. 1674), practitioner.
	James Halsy II (1695–1767); mathematical and surveying instruments.
	John Halsy (fl. 1700); mathematical instruments.
	Joseph Halsy (fl. 1697–1762); surveying instruments.
	Rowland Houghton (1678–1744); surveying instruments.
	Andrew Newell (1749–1798); surveying instruments.
	Joseph Newell (fl. 1800–1813); surveying instruments.
	Joseph Pope (1750–1826); scientific instruments and clocks.
	Nathan Prince (fl. 1743), practitioner; scientific instruments.
	Paul Revere (1735–1818); gunnery instruments.
	Charles Thacher (18th century); surveying instruments.
	Samuel Thaxter (1769–1842); surveying, nautical, and mathematical instruments.
	William Williams (1737/8–1792); mathematical and nautical instruments.
Brookfield:	John Potter (1746–1818); surveying instruments.
Hanover:	John Bailey II (1752–1823); surveying instruments.

Lanesboro:	Benjamin Platt (1757–1833); surveying instruments, clocks, and compasses.
Lynn:	John Bailey II (1752–1823); surveying instruments.
Nantucket:	Peter Folger (1617–1690), practitioner(?).
	Walter Folger (1765–1849), practitioner; clocks and astronomical instruments.
Newburyport:	Gideon Fairman (1774–1827); mathematical instruments.
	Hooker & Fairman (before 1810); mathematical instruments.
Northampton:	Jedidiah Baldwin (c.1777–1829); surveying instruments.
	Stiles & Baldwin (fl. 1791); surveying instruments.
	Stiles & Storrs (fl. 1792); surveying instruments.
Plymouth:	Benjamin Warren (fl. 1740–1790); surveying and nautical instruments.
Salem:	Jedidiah Baldwin (c.1777–1829); surveying instruments.
	Samuel Emery (1787–1882); mathematical instruments.
	John Jayne (late 18th century); mathematical instruments.
	Benjamin King (1740–1804); nautical instruments.
	Daniel King (1704–1790); mathematical instruments.
	John Prince (1751–1836), practitioner; scientific instruments.
Southwick:	Amasa Holcomb (1787–1875); surveying and mathematical instruments.

NEW HAMPSHIRE

Exeter:	Benjamin C. Gilman (1763–1835); mathematical instruments and clocks.
Hanover:	Jedidiah Baldwin (c.1777–1829); surveying instruments.
Newmarket:	John Kennard (1782–1861); surveying instruments.
Portsmouth:	Thomas S. Bowles (c.1765–1821); surveying instruments.
Walpole:	Gurdon Huntington (1763–1804); clocks and surveying and other instruments.

NEW JERSEY

Elizabeth:	Aaron Miller (fl. 1748–1771); surveying instruments, clocks, and compasses.
Wistarburg:	Richard Wistar (fl. 1752); glass and thermometric instruments.

NEW YORK

Fishkill:	John Bailey (fl. 1778); surveying and surgical instruments.
New York:	Thomas Biggs (fl. 1792); surveying instruments.
	Bulmain & Dennies (fl. 1799); nautical instruments.
	H. Caritat (fl. 1799); astronomical prints.
	John Donegan (fl. 1787); barometers, thermometers, and philosophical instruments.
	W. Fosbrook (fl. 1786); surgical and dental instruments.

New York—Continued	Joseph Gatty (fl. 1794); barometers, thermometers and philosophical instruments.
	James Ham (fl. 1754–1764); mathematical instruments.
	William Hinton (fl. 1772); mathematical instruments.
	A. Lamb & Son (fl. late 18th century); mathematical instruments.
	Anthony Lamb (1703–1784); mathematical instruments.
	John Lamb (1735–1800); mathematical instruments.
	M. Morris (fl. 1785); protractors.
	Widow Balthaser Sommer (fl. 1753); optical instruments.
	Charles Walpole (fl. 1746); mathematical instruments.
	James Youle (1740–1786); surgical instruments.
	John Youle (fl. 1786); surgical instruments.

OHIO

Columbus:	Augustus Platt (1793–1886); mathematical instruments.
	Benjamin Platt (1757–1833); surveying instruments and clocks.
Gallipolis:	Joseph (fl. 1792) and Francois Devacht; watches, compasses, and sundials.

PENNSYLVANIA

Bucks County:	W. L. Potts (late 18th century); surveying instruments.
	George Wall, Jr. (fl. 1788); surveying instruments.
Germantown:	Christopher Sower (c.1724–1740); mathematical instruments and clocks.
	Christopher Witt (fl. 1710–1765); mathematical instruments and clocks.
Harrisburg:	Frederick A. Heisely (1759–1839); clocks and mathematical instruments.
	George Heisely (1789–1880); clocks and mathematical instruments.
Lancaster:	George Ford (late 18th century to 1842); surveying and nautical instruments.
	George Ford II (fl. 1842); surveying and nautical instruments.
	Frederick A. Heisely (1759–1839); clocks and mathematical instruments.
	George Hoff (1740–1816); clocks, surveying instruments.
	Thomas Mendenhall (fl. 1775); mathematical instruments and clocks.
Norristown:	David Rittenhouse (1732–1796), practitioner; astronomical and surveying instruments.
Philadelphia:	Owen Biddle (1737–1799), practitioner.
	Thomas Biggs (fl. 1792–1795); surveying instruments.
	Isaac Brokaw (fl. 1771).
	Benjamin Condy (fl. 1756, d. 1798); mathematical instruments and sand glasses.

Philadelphia—Continued	William Davenport (1778–1829); surveying and mathematical instruments.
	William Dean (?–1797); surveying and nautical instruments.
	George Evans (fl. 1796, d. 1798); mathematical instruments.
	Martin Fisher (fl. 1790); glass instruments.
	Joseph Gatty (fl. 1794); barometers, thermometers, and philosophical instruments.
	Bryan Gilmur (end of 18th century); mathematical instruments and clocks.
	Thomas Godfrey (1704–1749); improved reflecting backstaff.
	John Gould (fl. 1794); nautical, surveying, and optical instruments.
	Theophilus Grew (fl. 1753), practitioner.
	James Ham (fl. 1754–1764); mathematical instruments.
	James Ham, Jr. (fl. 1780); mathematical instruments.
	Alloysius Ketterer (fl. 1789); glass instruments.
	Anthony Lamb (1703–1784); mathematical instruments.
	Thomas Pryor (fl. 1778); mathematical instruments.
	Benjamin Rittenhouse (1740–c.1820); surveying and astronomical instruments.
	David Rittenhouse (1732–1796), practitioner; astronomical and surveying instruments.
	Christopher Sower [Sauer] (c.1724–1740); mathematical instruments and clocks.
	Charles Taws (fl. 1795); mathematical instruments.
	Henry Voight (1738–1814); clocks, watches, and astronomical instruments.
	John Whitney (fl. 1801); mathematical and optical instruments.
	Thomas Whitney (fl. 1798–1823); mathematical and optical instruments.
	John Wood (fl. 1790); compasses.
Pittsburgh:	Frederick A. Heisely (1759–1839); clocks and mathematical instruments.
West Bradford:	Joel Baily (1732–1797), practitioner.

RHODE ISLAND

Newport:	William G. Hagger (c.1744–1830?); quadrants.
	King & Hagger (1759/60); mathematical and nautical instruments.
	Benjamin King (1707–1786); mathematical and nautical instruments.
	Samuel King (1748–1819); mathematical instrument.
	Paul Pease (fl. 1750); quadrants.
Providence:	William Hamlin (1772–1869); mathematical, astronomical, and nautical instruments.

SOUTH CAROLINA

Charleston: Charles Blundy (fl. 1753); thermometric instruments.
Robert Clark (fl. 1785); nautical, surveying, and optical instruments.
James Jacks (fl. 1780's); mathematical and surveying instruments.

VERMONT

Bradford: James Wilson (1763–1855); globes.

VIRGINIA

Winchester: Goldsmith Chandlee (c.1746–1821); surveying and astronomical instruments and clocks.
Anthony Lamb (1703–1784); mathematical instruments.

TYPES OF INSTRUMENTS AND THEIR MAKERS

(*Categories based on specific designations noted in advertisements*)

ASTRONOMICAL

Caritat, H. (fl. 1799), New York.

Chandlee, Goldsmith (c.1746–1821), Winchester, Va.; also made surveying instruments and clocks.

Ellicott, Andrew (1754–1820), Baltimore; also made surveying instruments.

Folger, Walter, Jr. (1765–1849), Nantucket, Mass.; also made surveying instruments.

Greenough, Thomas (1710–1785), Boston; also made mathematical, surveying, and nautical instruments.

Hamlin, William (1772–1869), Providence, R.I.; also made mathematical and nautical instruments.

Holcomb, Amasa (1787–1875), Southwick, Mass.; also made surveying instruments.

Rittenhouse, Benjamin (1740–c.1820), Philadelphia; also made surveying instruments.

Rittenhouse, David (1732–1796), Philadelphia and Norristown, Pa.; also made surveying instruments.

Voight, Henry (1738–1814), Philadelphia; also made clocks and watches.

GLASS AND THERMOMETRIC

Blundy, Charles (fl. 1753), Charleston, S.C.; also made watches.
Donegan, Joseph (fl. 1787), New York and Philadelphia; also made philosophical instruments.
Fisher, Martin (fl. 1790), Philadelphia.
Gatty, Joseph (fl. 1794), New York and Philadelphia; also made philosophical instruments.
Ketterer, Alloysius (fl. 1789), Philadelphia.
Wistar, Richard (fl. 1752), Wistarburg, N.J.

HOROLOGICAL

Blundy, Charles (fl. 1753), Charleston, S.C.; also made thermometric instruments.
Burnap, Daniel (1759–1838), East Windsor and Coventry, Conn.; also made surveying instruments.
Chandlee, Benjamin (1723–1791), Nottingham, Md.; also made surveying instruments.
Chandlee & Bros. (fl. 1790–1791), Nottingham, Md.; also made surveying instruments.
Chandlee, Ellis (1755–1816), Nottingham, Md.; also made surveying instruments.
Chandlee, Goldsmith (1751–1821), Winchester, Va.; also made astronomical and surveying instruments.
Chandlee, Isaac (1760–1813), Nottingham, Md.; also made surveying instruments.
Crow, George (c. 1726–1772), Philadelphia; also made surveying instruments.
DeVacht, Joseph and Francois (fl. 1792), Gallipolis, Ohio; also made compasses and sundials.
Doolittle, Enos (1751–1806), Hartford, Conn.; also made surveying and nautical instruments.
Doolittle, Isaac (1721–1800), New Haven, Conn.; also made scientific instruments.
Doolittle, Isaac Jr. (1759–1821), New Haven, Conn.; also made surveying instruments.
Gilman, Benjamin C. (1763–1835), Exeter, N.H.; also made mathematical instruments.
Gilmur, Bryan (fl. end of 18th century), Philadelphia; also made mathematical instruments.
Harland, Thomas (1735–1807), Norwich, Conn.; also made surveying instruments.
Heisely, Federick A. (1759–1839), Frederick, Md.; also made mathematical and surveying instruments.
Heisely, George (1789–1880), Harrisburg, Pa.; also made mathematical instruments.
Hoff, George (1740–1816), Lancaster, Pa.; also made surveying instruments.
Huntington, Gurdon (1763–1804), Windham, Conn., and Walpole, N.H.; also made surveying and other instruments.
Kennard, John (1782–1861), Newmarket, N.H.; also made surveying instruments.
Mendenhall, Thomas (fl. 1775), Lancaster, Pa.; also made mathematical instruments.
Miller, Aaron (fl. 1748–1771), Elizabethtown, N.J.; also made compasses and surveying instruments.

Platt, Benjamin (1757–1833), Danbury, Litchfield, and New Milford, Conn.; Lanesboro, Mass.; Columbus, Ohio; also made compasses and surveying instruments.

Pope, Joseph (1750–1826), Boston; also made scientific instruments.

Sibley & Marble (Clark Sibley and Simeon Marble) (late 18th century), New Haven, Conn.; also made mathematical instruments.

Sower, Christopher (c. 1724–1740), Germantown and Philadelphia, Pa.; also made mathematical instruments.

Voigt, Henry (1738–1814), Philadelphia; also made astronomical instruments.

White, Peregrine (1747–1834), Woodstock, Conn.; also made surveying instruments.

Witt, Christopher (practitioner) (fl. 1710–1765), Germantown, Pa.; also made mathematical instruments.

MATHEMATICAL (GENERAL)

Condy, Benjamin (fl. 1756–1792, d. 1798), Philadelphia.

Dabney, John, Jr. (fl. 1739), Boston.

Dakin, Jonathan (fl. 1745), Boston; also made balances.

Davenport, William (fl. 1800–1820), Philadelphia; also made surveying instruments.

Doolittle, Isaac (1721–1800), New Haven, Conn.; also made clocks.

Emery, Samuel (late 18th century), Salem, Mass.

Evans, George (fl. 1796, d. 1798), Philadelphia.

Fairman, Gideon (1774–1827), Newburyport, Mass.

Gilman, Benjamin C. (1763–1835), Exeter, N.H.; also made clocks.

Gilmur, Bryan (end of 18th century), Philadelphia; also made clocks.

Greenleaf, Stephen (fl. 1745), Boston.

Greenough, Thomas (1710–1785), Boston; also made surveying, astronomical, and nautical instruments.

Greenwood, Isaac, Jr. (1730–1803), Boston.

Hagger, Benjamin K. (c.1769–1834), Boston and Baltimore; also made surveying instruments.

Halsy, James, II (1695–1767), Boston; also made surveying instruments.

Halsy, John (fl. 1700), Boston.

Ham, James (fl. 1754–1764), New York and Philadelphia.

Ham, James, Jr. (fl. 1780), Philadelphia.

Hamlin, William (1772–1869), Providence, R.I.; also made nautical and astronomical instruments.

Heisely, Frederick (1759–1839), Frederick, Md., and Lancaster, Harrisburg, and Pittsburgh, Pa.; also made clocks and surveying instruments.

Heisely, George (1789–1880), Harrisburg, Pa.; also made clocks.

Hinton, William (fl. 1772), New York.

Hooker & Fairman (before 1810), Newburyport, Mass.

Jacks, James (fl. 1780's), Charleston, S.C.; also made surveying instruments.

Jayne, John (late 18th century), Salem, Mass.

King & Hagger (1759/60 to early 1760's), Newport, R.I.; also made nautical instruments.

King, Benjamin (1707–1786), Newport, R.I.; also made nautical instruments.

King, Daniel (1704–1790), Salem, Mass.
King, Samuel (fl. 1786), Newport, R.I.
Lamb, A. & Son (1780's), New York.
Lamb, Anthony (1703–1784), Virginia, Philadelphia, New York, and Hunter's Key, N.Y.; also made surveying and nautical instruments.
Lamb, John (1735–1800), New York; also made nautical and surveying instruments.
Mendenhall, Thomas (fl. 1775), Lancaster, Pa.; also made clocks.
Newell, Andrew (1749–1798), Boston; also made compasses and surveying instruments.
Platt, Augustus (1809–1886), Columbus, Ohio; also made surveying instruments.
Pryor, Thomas (fl. 1778), Philadelphia.
Revere, Paul (1735–1818), Boston, Mass.
Sibley & Marble (late 18th century), New Haven, Conn.; also made clocks and watches.
Sower, Christopher (c. 1724–1740), Germantown and Philadelphia, Pa.; also made clocks.
Taws, Charles (fl. 1795), Philadelphia.
Thaxter, Samuel (1769–1842), Boston; also made surveying and nautical instruments.
Walpole, Charles (fl. 1746), New York.
Whitney, John (fl. 1801), Philadelphia; also made optical instruments.
Whitney, Thomas (fl. 1798–1821), Philadelphia; also made optical and surveying instruments.
Williams, William (1737/38–1792), Boston; also made nautical instruments.
Witt, Christopher (fl. 1710–1765), Germantown, Pa.; also made clocks.

NAUTICAL

Bulmain & Dennies (fl. 1799), New York.
Clark, Robert (fl. 1785), Charleston, S.C.; also made surveying and optical instruments.
Condy, Benjamin (fl. 1756–92, d. 1798), Philadelphia; also made mathematical instruments.
Davenport, William (fl. 1800–1820), Philadelphia; also made mathematical and surveying instruments.
Dean, William (?–1797), Philadelphia; also made surveying instruments.
Doolittle, Enos (1751–1806), Hartford, Conn.; also made surveying instruments, directional compasses and clocks.
Emery, Samuel (1787–1882), Salem, Mass.
Fairman, Gideon (1774–1827), Newburyport, Mass.; also made mathematical instruments.
Ford, George, I (late 18th century to 1840), Lancaster, Pa.; also made surveying instruments.
Ford, George, II (fl. 1842), Lancaster, Pa.; also made surveying instruments.
Godfrey, Thomas (1704–1749), Philadelphia.
Gould, John (fl. 1794), Philadelphia; also made surveying and optical instruments.

Greenough, Thomas (1710–1785), Boston; also made mathematical and surveying instruments.

Hagger, William G. (c.1744–1830?), Newport, R.I.

Ham, James (fl. 1754–64), New York and Philadelphia; also made mathematical instruments.

Ham, James, Jr. (fl. 1780), Philadelphia; also made mathematical instruments.

Hamlin, William (1772–1869), Providence, R.I.; also made mathematical instruments.

Jayne, John (late 18th century), Salem, Mass.; also made mathematical instruments.

King & Hagger (1759/60 to early 1760's), Newport, R.I.; also made mathematical instruments.

King, Benjamin (1707–1786), Newport, R.I.; also made mathematical instruments.

King, Benjamin (1740–1804), Salem, Mass.

King, Daniel (1704–1790), Salem, Mass.; also made mathematical instruments.

King, Samuel (fl. 1786), Newport, R.I.; also made mathematical instruments.

Lamb, A., & Son (1780's), New York; also made mathematical instruments.

Lamb, Anthony (1703–1784), Virginia, Philadelphia, New York, and Hunter's Key, N.Y.; also made mathematical and surveying instruments.

Lamb, John (1735–1800), New York; also made surveying and mathematical instruments.

Newell, Andrew (1749–1798), Boston; also made mathematical instruments.

Pease, Paul (fl. 1750), probably Rhode Island.

Thaxter, Samuel (1769–1842), Boston; also made mathematical and surveying instruments.

Warren, Benjamin (fl. 1740–1790), Plymouth, Mass.; also made surveying instruments.

Williams, William (1737/38–1792), Boston; also made mathematical instruments.

OPTICAL

Benson, John (fl. 1793–1797).

Clark, Robert (fl. 1785), Charleston, S.C.; also made nautical and surveying instruments.

Sommer, Widow Balthaser (fl. 1753), New York.

Whitney, John (fl. 1801), Philadelphia; also made mathematical instruments.

Whitney, Thomas (fl. 1798–1821), Philadelphia; also made mathematical and surveying instruments.

SURGICAL

Bailey, John (fl. 1778), Fishkill, N.Y.; also made surveying instruments.

Fosbrook, W. (fl. 1786), New York; also made dental instruments.

Youle, James (1740–1786), New York.

Youle, John (fl. 1786), New York.

SURVEYING

Bailey, John (fl. 1778), Fishkill, N.Y.; also made surgical instruments.
Bailey, John, II (1752–1823), Hanover and Lynn, Mass.
Baldwin, Jedidiah (c.1777–1829), Salem, Boston, and Northampton, Mass., and Hanover, N.H.
Biggs, Thomas (fl. 1792–1795), New York and Philadelphia.
Blakeslee, Ziba (1768–1834), Newtown, Conn.
Bowles, Thomas S. (c.1765–1821?), Portsmouth, N.H.
Breed, Aaron (late 18th to mid-19th centuries), Boston.
Burnap, Daniel (1759–1838), East Windsor and Coventry, Conn.; also made clocks.
Chandlee, Benjamin, Jr. (1723–1791), Nottingham, Md.; also made clocks.
Chandlee & Bros. (fl. 1790–1791), Nottingham, Md.; also made clocks.
Chandlee, Ellis (1755–1816), Nottingham, Md.; also made clocks.
Chandlee, Ellis & Bros. (fl. 1791–1797), Nottingham, Md.; also made clocks.
Chandlee, Goldsmith (c.1746–1821), Winchester, Va.; also made clocks and sundials.
Chandlee, Isaac (1760–1813), Nottingham, Md.; also made clocks.
Clark, Robert (fl. 1785), Charleston, S.C.; also made nautical and optical instruments.
Clough, Jere (18th century), Boston.
Crow, George (fl. 1754–1772), Wilmington, Del.; also made clocks.
Davenport, William (fl. 1800–1820), Philadelphia; also made mathematical instruments.
Dean, William (?–1797), Philadelphia; also made nautical instruments.
Doolittle, Enos (1751–1806), Hartford, Conn.; also made nautical instruments and clocks.
Doolittle, Isaac, Jr. (1759–1821), New Haven, Conn.; also made clocks.
Dupee, John (after 1761), Boston.
Ellicott, Andrew (1754–1820), Baltimore; also made astronomical instruments.
Ford, George, I (late 18th century to 1840), Lancaster, Pa.; also made nautical instruments.
Ford, George, II (fl. 1842), Lancaster, Pa.; also made nautical instruments.
Gould, John (fl. 1794), Philadelphia; also made nautical and optical instruments.
Greenough, Thomas (1710–1785), Boston, also made nautical and mathematical instruments.
Greenough, William (fl. 1785), Boston.
Halsy, James, II (1695–1767), Boston; also made mathematical instruments.
Halsy, Joseph (fl. 1697–1762), Boston.
Hagger, Benjamin K. (c. 1769–1834), Boston and Baltimore; also made mathematical instruments.
Hanks, Benjamin (1755–1824), Mansfield and Litchfield, Conn.
Hanks, Truman (fl. 1808), Mansfield and Litchfield, Conn.
Harland, Thomas (1735–1807), Norwich, Conn.; also made clocks.
Heisely, Frederick A. (1759–1839), Frederick, Md., and Lancaster, Harrisburg, and Pittsburgh, Pa.; also made clocks and mathematical instruments.
Heisely, George (1789–1880), Harrisburg, Pa.; also made clocks and mathematical instruments.

Holcomb, Amasa (1785–1875), Southwick, Mass.; also made astronomical instruments.

Houghton, Rowland (c. 1678–1744), Boston.

Huntington, Gurdon (1763–1804), Windham, Conn., and Walpole, N.H.; also made clocks and other scientific instruments.

Jacks, James (fl. 1780's), Charleston, S.C.; also made mathematical instruments.

Kennard, John (1782–1861), Newmarket, N.H.; also made clocks.

Lamb, A., & Son (1780's), New York; also made mathematical and nautical instruments.

Lamb, Anthony (1703–1784), New York; also made mathematical and nautical instruments.

Lamb, John (1735–1800), New York; also made mathematical and nautical instruments.

Miller, Aaron (fl. 1748–1771), Elizabeth, N.J.; also made clocks and directional compasses.

Newell, Andrew (1749–1798), Boston; also made mathematical instruments and directional compasses.

Platt, Augustus (1809–1886), Columbus, Ohio; also made mathematical and surveying instruments.

Platt, Benjamin (1757–1833), Danbury, Litchfield, and New Milford, Conn.; Lanesboro, Mass.; and Columbus, Ohio; also made directional compasses and clocks.

Potter, John (fl. 1785), Brookfield, Mass.

Rittenhouse, Benjamin (1740–c. 1820), Philadelphia; also made astronomical instruments.

Rittenhouse, David (1732–1796), Philadelphia; also made astronomical instruments.

Rittenhouse & Evans (fl. 1770's), Philadelphia.

Stiles & Baldwin (fl. 1791), Northampton, Mass.

Stiles & Storrs (fl. 1792), Northampton, Mass.

Thacher, Charles, probably Boston.

Thaxter, Samuel (1769–1842), Boston; also made nautical and mathematical instruments.

Wall, George Jr. (fl. 1788), Bucks County, Pa.

Warren, Benjamin (fl. 1740–1790), Plymouth, Mass.; also made nautical instruments.

White, Peregrine (1747–1834), Woodstock, Conn.; also made clocks.

Whitney, Thomas (fl. 1798–1821), Philadelphia; also made mathematical and optical instruments.

Williams, William (1737/38–1792), Boston; also made nautical instruments.

Bibliography of Published Sources

ADAMS, GEORGE. *Mathematical and geographical essays.* London, 1791.

ABBOTT, KATHERINE M. *Old paths and legends of New England.* New York: G. P. Putnam's Sons, 1909.

BABB, MAURICE J. David Rittenhouse. *The Pennsylvania Magazine of History and Biography* (July 1932), vol. 56, no. 223, pp. 193–224.

BARTON, WILLIAM. *Memoirs of the life of David Rittenhouse, L.L.D., F.R.S.* Philadelphia, 1813.

BEDINI, SILVIO A. A compass card by Paul Revere (?). *Yale Library Gazette* (July 1962), vol. 37, no. 1, pp. 36–38.

BEDINI, SILVIO A. *Ridgefield in review.* New Haven: Walker-Rackliffe Co., 1958.

BENTLEY, WILLIAM. *The diary of William Bentley, D.D.* Salem, Mass., 1905.

BION, NICOLAS. *Traitè de la construction et dès principaux usages des instruments de mathematiques.* Paris, 1709. Transl. Edmund Stone, London, 1724.

BRANCH, W. J. V., and BROOK-WILLIAMS, Capt. E. *A short history of navigation.* Annapolis, Md.: Weems System of Navigation, 1942.

BREWSTER, CHARLES W. *Rambles about Portsmouth.* Ser. 1. Portsmouth, N.H.: L. W. Brewster, 1859.

———. *Rambles about Portsmouth.* Ser. 2. Portsmouth, N.H.: L. W. Brewster, 1869.

BRIDENBAUGH, CARL. *The colonial craftsman.* New York: N.Y. University Press, 1950.

——— and BRIDENBAUGH, J. *Rebels and gentlemen: Philadelphia in the age of Franklin.* New York: Reynals and Hitchcock, 1942.

BRIGHAM, CLARENCE S. *Paul Revere's engravings.* Worcester, Mass.: American Antiquarian Society, 1954.

CAJORI, F. *The teaching and history of mathematics in the United States.* (Bureau of Education Circular of Information 3.) Washington: Bureau of Education, 1890.

———. *The early mathematical sciences in North and South America.* Boston: Badger, 1928.

CHANDLEE, EDWARD E. *Six Quaker clockmakers.* Philadephia: Pennsylvania Historical Society, 1943.

CHAPIN, HOWARD M. Davis quadrants. *Antiques* (November 1927), vol. 12, no. 5, pp. 397–399.

CONRAD, HENRY C. Old Delaware clockmakers. *The Historical and Biographical Papers of the Historical Society of Delaware* (1897), vol. 3, chapt. 20.

COHEN, I. BERNARD. *Some early tools of American science.* Cambridge: Harvard University Press, 1950.

DAVIS, H. S. David Rittenhouse. *Popular Astronomy* (July 1896), vol. 4, no. 1, pp. 1–12.

DAVIS, WILLIAM T. *Ancient landmarks of Plymouth.* Boston: A. Williams & Co., 1883.

DAY, J. *Principles of navigation and surveying.* New Haven, Conn., 1817.

DOW, GEORGE FRANCIS. *The arts and crafts in New England 1704–1775.* Topsfield, Mass.: The Wayside Press, 1927.

DYER, WALTER A. *Early American craftsmen.* New York: Century Co., 1915.

ECKHARDT, GEORGE H. *Pennsylvania clocks and clockmakers.* New York: Devin-Adair Co., 1955.

ELLICOTT, ANDREW. *The journal of Andrew Ellicott, late Commissioner on behalf of the United States during part of the year 1796, the years 1797, 1798, 1799, and part of the year 1800 for determining the boundary between the United States and the possessions of his Catholic Majesty in America.* Philadelphia: Budd and Barton, 1803.

EVANS, GEORGE. *Illustrated history of the United States mint.* Philadelphia: Evans, 1890.

FELT, JOSEPH B. *Annals of Salem.* Salem, Mass., 1827.

FITTS, Rev. JAMES HILL. *History of Newfields, New Hampshire, 1638–1911.* Concord: Rumford Press, 1912.

FLINT, ABEL. *System of geometry and trigonometry, together with a treatise of surveying.* Hartford: Olive D. Cook, 1804.

FORBES, ESTHER. *Paul Revere and the world he lived in.* Boston: Houghton, Mifflin Co., 1942.

Frederick A. Heisely, watch and clockmaker and his recorded years, 1759–1839. *Timepieces Quarterly* (November 1948), vol. 1, no. 1, p. 33.

GARDNER, WILL, *The clock that talks and what it tells.* Nantucket: Nantucket Whaling Museum, 1954.

GILLINGHAM, HARROLD E. Some early Philadelphia instrument makers. *The Pennsylvania Magazine of History and Biography* (1927), vol. 51, no. 3, pp. 289–308.

———. The first orreries in America. *Journal of the Franklin Institute* (1940), vol. 229, pp. 81–99.

GOTTESMAN, RITA, *The arts and crafts in New York, 1726–1776.* New York: N. Y. Historical Society, 1938.

———. *The arts and crafts in New York, 1777–1799.* New York: N. Y. Historical Society, 1954.

GREENWOOD, ISAAC J. *The Greenwood family.* Privately printed, 1934.

HAMILTON, ALEXANDER. *Official reports on publick credit, a national bank, manufactures and a mint.* Philadelphia: Wm. McKean, 1821.

HINDLE, BROOKE. *The pursuit of science in revolutionary America 1735–1789.* Chapel Hill: University of North Carolina Press, 1956.

History of Hingham, Mass. Hingham, 1893.

HOCKER, EDWARD W. *A doctor of colonial Germantown, Christopher Witt, physician, mystic and seeker after the truth.* Germantown, Pa.: Germantown Historical Society, 1948.

HOOPES, PENROSE R. *Connecticut clockmakers of the eighteenth century.* New York: Dodd Mead & Co., 1930.

———. *Early clockmaking in Connecticut.* New Haven: Yale University Press, 1934.

———. *Shop records of Daniel Burnap, clockmaker.* Hartford, Conn.: Connecticut Historical Society, 1958.

HUNTER, FREDERICK W. *Stiegel glass.* Boston: Houghton Mifflin Co., 1914.

[Huntington], *Memoirs of the Huntington Family Association*, Hartford, Conn.: privately printed, 1915.

JAFFE, BERNARD. *Men of science in America.* New York: Simon & Schuster, 1944.

JAMES, ARTHUR E. *Chester County clocks and their makers.* West Chester, Pa., 1947.

KARPINSKI, L. C. *Bibliography of mathematical works printed in America through 1850.* Ann Arbor, Mich.: University of Michigan Press, 1940.

KIELY, EDMOND R. *Surveying instruments, their history and classroom use.* New York: Teachers College, Columbia University, 1947.

KIMBALL, LEROY E. James Wilson of Vermont, America's first globe maker. *Proceedings of the American Antiquarian Society* (April 1938), new ser., vol. 48, no. 1, pp. 29–48.

KING, RUFUS. *Pedigree of King of Lynn.* Salem, Mass., 1891.

KINGMAN, E. D. Roger Sherman, colonial surveyor. *Civil Engineering* (August 1940), vol. 10, no. 8, pp. 514–515.

LANE, GLADYS R. Rhode Island's earliest engraver. *Antiques* (March 1925), vol. 7, no. 3, pp. 133–137.

LATROBE, JOHN H. B. Memoir of Benjamin Banneker. *Maryland Colonization Journal* (May 1845).

LEAKE, ISAAC Q. *Memoir of the life and times of General John Lamb.* Albany: Munsell, 1850.

LEPHILLIPS, PHILIP. The Negro, Benjamin Benneker. *Records of the Columbia Historical Society* (1916), vol. 20, pp. 114–120.

LEYBOURN, WILLIAM. *The compleat surveyor.* London, 1653.

LOVE, JOHN. *Geodasia, or the art of surveying.* London, 1688.

LOWNES, A. E. The 1769 transit of Venus and its relation to early American astronomy. *Sky and Telescope* (1943), vol. 2.

MAGEE, D. F. Grandfather's clocks: Their making and their makers in Lancaster County. Paper read before the Lancaster (Pa.) Historical Society, 1917.

MATHEWS, CATHERINE VAN CORTLANDT. *Andrew Ellicott, his life and letters.* New York: Grafton Press, 1908.

McCABE, WILLIAM. Benjamin Platt of New Fairfield, Connecticut. *Timepieces Quarterly* (November 1948), vol. 1, no. 1, pp. 26–29.

MILHAM, WILLIS I. Early American observatories. *Popular Astronomy* (November and December 1937), vol. 14, nos. 9 and 10.

———. *The history of astronomy in Williams College and the founding of Hopkins Observatory.* Williamstown, Mass.: Williams College, 1937.

———. *Early American observatories: Which was the first astronomical observatory in America?* Williamstown, Mass.: Williams College, 1938.

MITCHELL, EDWIN VALENTINE. *The romance of New England antiques.* New York, A. A. Wyn, 1950.

MOORE, S. *An accurate system of surveying.* Litchfield, Conn.: T. Collier, 1796.

MULTHAUF, ROBERT P. Early instruments in the history of surveying: Their use and invention. *Surveying and Mapping* (October–December, 1958), pp. 399–415.

———. ed. Holcomb, Fitz and Peate, three 19th-century American telescope makers. Paper 26 in *Contributions from the Museum of History and Technology Papers 19–30* (U.S. National Museum Bulletin 228), Washington: Smithsonian Institution, 1962.

PALMER, BROOKS. *The book of American clocks.* New York: Macmillan Co., 1950.

PHILLIPS, JOHN M. An unrecorded engraving by Nathaniel Hurd. *Bulletin of the Associates in Fine Arts at Yale University* (June 1936), vol. 7, no. 2, pp. 26–27.

PRICE, DEREK J. DE SOLLA. *Science since Babylon.* New Haven: Yale University Press, 1961.

PRIME, ALFRED COXE. *The arts and crafts of Philadelphia, Maryland and South Carolina, 1721–1785.* Ser. 1. Topsfield, Mass.: Walpole Society, 1929.

———. *The arts and crafts of Philadelphia, Maryland and South Carolina, 1786–1800.* Ser. 2. Topsfield, Mass.: Walpole Society, 1929.

RATHBORNE, AARON. *The surveyor; in four bookes.* London: W. Standsby, 1616.

RAYNER, W. H. From Columbus' compass to the first transit. *Civil Engineering* (1939), vol. 9, no. 11, pp. 661–664.

Report of the Committee on the Rooms. *Proceedings of the Bostonian Society* (1917), vol. 40, no. 1, pp. 14–16.

SAVAGE, JAMES. *A genealogical dictionary of the first settlers of New England.* 2 vols. Boston, 1860.

SCHOEN, H. H. The making of maps and charts. In *Ninth Yearbook of the Council for Social Studies.* Cambridge: Harvard University Press, 1938.

SEYBOLD, R. F. The evening school in colonial America. *University of Illinois Bureau of Educational Research,* Bulletin 31. 1925.

STEELE, A. P. *The history of Clark County, Ohio.* Chicago: W. H. Beers Co., 1881.

STEVENSON, D. ALAN. *The world's lighthouses before 1820.* London: Oxford University Press, 1959.

STRETCH, CAROLYN WOOD. Early colonial clockmakers in Philadelphia. *Pennsylvania Magazine of History and Biography* (July 1932), vol. 56, no. 223, p. 666.

STRUIK, DIRK J. *Yankee science in the making.* Boston: Little Brown & Co., 1948.

TAYLOR, E. G. R. *The mathematical practitioners of Tudor and Stuart England.* Cambridge University Press, 1954.

THOMPSON, SYLVANUS. The rose of the winds. *Proceedings of the British Academy, 1913–14, 10th Annual Conference,* pp. 179–211.

UPHAM, C. W. Memoir of the Reverend John Prince. *American Journal of Science* (1837), vol. 31, pp. 201–222.

WHITTLESEY, C. Origin of the American system of land surveys. *Journal of the Association of Engineering Societies* (July 1883), vol. 3.

WIENBERGER, BERNARD W. *Introduction to the history of dentistry.* St. Louis: Mosby Co., 1948.

Index

ERRATA AND ADDENDA

Page			
6	...par. 2	.line 5	.. "issues" not "issue"
12	...Fig. 4	1	.. "Rittenhouse and Potts"
15	...par. 1	3	.. "Pennsylvania" not "Philadelphia"
15	 1	 8	.. "same" not "following"
16	 7	 6	.. "made" not "make"
19	...par. 1	 8	.. "1790" not "1780"
19	3	7	.. "Apalachiola" not "Appalachiola"
20	..Fig. 10	....	Replace with Fig. 31 on page 64.
22	... par. 1	. 6 - 7	.. Delete "he acquired a watch from a trader" and add "Tradition states he saw a watch owned by a trader"
22	1	.10-11	.. Delete "Banneker . . . and"
22	1	...13	.. Delete "Joseph" add "the Ellicotts"
22	1	19-20	.. Delete "three years" add "a short time"
22	1	.22-24	.. Delete "during to "region."
28	...par. 4	2	.. Anthony Lamb (1703-1784)
29	....	.4-10	.. Delete "After" to "prospered." Add "Nothing is known of his activities thereafter until he established himself in New York in 1730."
29	....	..12	.. Delete "first"
37	2	2	.. Stephen Greenleaf (1704-1795)
38	2	1	.. Rowland Houghton (c. 1678-1744)
38	3	2	.. He was related to Thomas Grenough and may have served an apprenticeship with him.
43	1	1	.. Samuel Emery (1787-1882)
43	2	1	.. John Potter (1746-1818)
43	4	3	.. Samuel King (1748-1819)
47	4	1	.. "Newtown" not "Newton"
51	1	4	.. Fredericksburg, "New York" not "Va."
53	3	2	.. Aaron Miller (fl. 1748-1771)
57	1	...10	.. made "in 1793"
61	2	...1-3	.. Tawes was a builder of pianos, not scientific instruments. Included in error.
62	.. par. 1	....1.1	. . "Voigt" not "Voight."
	2	1	. . "Voigt" not "Voight"
	2	4	. . "made in 1804"
63	2	2	.. "1842" not "1840"
63	6	1	.. James Jack (fl. 1780-1797)
64	..Fig. 31	....	Replace with Fig. 10 from p. 20

65 . . .par. 4 . . . .2 . . James Ham (fl. 1754-1764)
73 . . . . . . .1 . . . .4 . . "half" not "quarter"
80 . . . . . . .2 . . . .2 . . Joseph Halsy (fl. 1697-1762)
81 . . . . . . .2 . . . .6 . . "on" not "in"
124 . . . . . . .3 . . . .1 . . Thomas S. Bowles (1785–)
125 . . . . . . .3 . . . .3 . . "Emery" not "Emerly"
129 . . . . . . .1 . . . .2 . . John Kennard (1782-1861)
145 . .Fig. 79 . . . .4 . . "Duvall" not "Duyal"
145 . . . . . .79 . . . .6 . . "Duvall" not "Duyal"
145 . . . . . .79 . . . .6 . . Kimmel ?????
150 . . .par. 4 . . . .6 . . "Culpeper" not "Culpepper"
155 . . . . Chandlee, Ellis & Bros. (fl. 1790-1797)
155 . . . . Chandlee, Goldsmith (1746-1821)
155 . . . . Chandlee & Sons (-1790)
155 . . . . Chandlee, John (fl. 1797)
156 . . . . Condy, Benjamin (fl. 1756-1792)
157 . . . . Greenwood, Isaac, Sr. (c. 1725-1750)
157 . . . . Halsy, John (fl. 1700-1750)
158 . . . . Jacks, James (fl. 1780-1797)
159 . . . . Sower, Christopher (1693-1740)
161 . . . . Rowland Houghton (c. 1678-1744)
163 . . . . A. Lamb & Son (fl. 1756)
163 . . . . Christopher Sower (1693-1740)
164 . . . . Christopher Sower (1693-1740)
165 . . . . James Jacks (fl. 1780-1797)
166 . . . . Heisley, Frederick A.
167 . . . . Sower, Christopher (1693-1740)
167 . . . . Davenport, William (1778-1829)
167 . . . . Emery, Samuel (1787-1882)
167 . . . . Halsy, John (fl. 1700-1750)
167 . . . . Jacks, James (fl. 1780-1797)
168 . . . . King, Samuel (1748-1819)
168 . . . . Platt, Augustus (1793-1886)
168 . . . . Sower, Christopher (1693-1740)
168 . . . . Ford, George I (late eighteenth century to 1842)
169 . . . . King, Samuel (1748-1819)
170 . . . . Crow, George (fl. c. 1726-1772)
170 . . . . Davenport, William (1778-1829)
170 . . . . Dean, William (fl. 1792-1797)
170 . . . . Ford, George I (late eighteenth century to 1842)
171 . . . . Holcomb, Amasa (1787-1875)
171 . . . . Jacks, James (fl. 1780-1797)
171 . . . . Platt, Augustus (1793-1886)

ADDENDA TO BIBLIOGRAPHY

BEDINI, Silvio A. *A Compass Card by Paul Revere?* Yale University Library Gazette, vol. 37. No. 1, July 1962, pp. 36-38.

BEDINI, Silvio A. *The Life of Benjamin Banneker*. New York: Charles Scribner's Sons, 1972; Rancho Cordova, Cal.: Landmark Enterprises, 1982.

BEDINI, Silvio A. *Thinkers and Tinkers. Early American Men of Science*. New York: Charles Scribner's Sons, 1975; Rancho Cordova, Cal.: Landmark Enterprises, 1983.

BEDINI, Silvio A. *Artisans in Wood. The Mathematical Instrument Makers*. In America's Wooden Age. Aspects of Its Early Technology. Edited by Brooke Hindle. Tarrytown, N. Y.: Sleepy Hollow Restorations, 1975, pp. 85-119.

BEDINI, Silvio A. *Andrew Ellicott: Surveyor of the Wilderness*. Surveying and Mapping, vol. XXXVI, No. 2, June 1976, pp. 113-35. (Contained in *Plotters and Patterns*. Landmark Publication number 974.

BEDINI, Silvio A. *The Practical Sciences in the American Revolution*. The Daughters of the American Revolution Magazine, vol. 110, No. 6, July 1976, pp. 766-75.

BEDINI, Silvio A. *Early American Surveyors. Mapping the Wilderness*. The Professional Surveyor, vol. 2, No. 2, March-April 1982, pp. 16-18.

BEDINI, Silvio A. *The Scientific Instruments of the Lewis and Clark Expedition*. Great Plains Quarterly vol. 4, No. 1. Winter 1984, pp. 54-69. (Available from Landmark Enterprises.)

BEDINI, Silvio A. *At The Sign of the Compass and Quadrant*. The Life and Times of Anthony Lamb. Transactions of the American Philosophical Society, vol. 74, Part 1, 1984. (Available from Landmark Enterprises.)

BREWINGTON, M. V. *The Peabody Museum Collection of Navigating Instruments With Notes on Their Makers*. Salem, Mass.: Peabody Museum 1963.

BROWN, Leland A. *Early Philosophical Apparatus at Transylvania College (And Relics of the Medical Department)*. Lexington, Ky.: Transylvania College Press, 1959.

LAGEMANN, Robert T. *The Garland Collection of Classical Physics Apparatus at Vanderbilt University*. Nashville, Tenn.: Folio Publishers, 1983.

W. E. Knowles Middleton. *Catalog of Meteorological Instruments in the Museum of History and Technology*. Smithsonian Studies in History and Technology No. 2 Washington, D. C.: Smithsonian Institution Press, 1969.

MULTHAUF, Robert P. Assisted by David Davies. *Catalogue of Instruments and Models in the Possession of the American Philosophical Society*. Memoirs vol. 53. Philadelphia: American Philosophical Society, 1961.

SMART, Charles E. *The Makers of Surveying Instruments in America Since 1700*. Vol. I. Troy, N. Y.: Regal Art Press, 1962. vol. II. 1967.

THOMSON, Don W. *Men and Meridians. The History of Surveying and Mapping in Canada*. Vol. I. Prior to 1867. Vol. 2. 1867 to 1917. Ottawa: Queen's Printer, 1966.

WHEATLAND, David P. Assisted by Barbara Carson. *The Apparatus of Science at Harvard 1765-1800. Cambridge, Mass.: Harvard University Press, 1968.*

No. 3035 Price 30 cts